Lecture Notes in Mathematics

Edited by A. Dold and B. Eckmann

473

Dan Burghelea
Richard Lashof
Melvin Rothenberg
(with an appendix by E. Pedersen)

Groups of Automorphisms of Manifolds

Springer-Verlag
Berlin · Heidelberg · New York 1975

Authors
Prof. Dan Burghelea
Mathematical Institute
of the Roumanian Academy
Bucharest/Roumania

Prof. Richard Lashof
Prof. Melvin Rothenberg
University of Chicago
Dept. of Mathematics
5734 University Avenue
Chicago, Illinois 60637
USA

AMS Subject Classifications (1970): 57A99, 57C35, 57C50, 57D05, 57D10, 57D40, 57D50, 57D65, 57E05, 57F10, 58D05

ISBN 3-540-07182-2 Springer-Verlag Berlin · Heidelberg · New York
ISBN 0-387-07182-2 Springer-Verlag New York · Heidelberg · Berlin

Printed in Germany
Offsetdruck: Julius Beltz, Hemsbach/Bergstr.

CONTENTS

Chapter 1. **Morlet's Lemma of Disjunction** 1

Chapter 2. **Lemma of Disjunction - 2nd Form** 20

Chapter 3. **Automorphisms and Concordances** 28

Chapter 4. **Fibrations over Spheres** 45

Chapter 5. **Fibrations over Manifolds** 55

Chapter 6. **The Structure of $A(M \times S^1)$** 102

Appendix I. **Kan Sets of Embeddings and Automorphisms** 110

Appendix II. **The Topological Category** 142

Bibliography 153

INTRODUCTION

This set of notes presents some of the results of the authors and others in the study of homotopy properties of groups of automorphisms of manifolds. The subject is old, but modern developments in differential and geometric topology have made possible dramatic new advances along a broad front, and have opened up many new and exciting problems. We have not, in these notes, attempted to survey all these new developments, but rather to go more deeply into certain questions which are of interest to us and to demonstrate the power and potential of the technique.

This work falls into three parts. The first three chapters investigate the relationship between the homotopy properties of a manifold M, and the homotopy properties of certain interesting groups of automorphisms of M and $M \times I$. The basic geometric result which underlies this work is Morlet's lemma of disjunction. In fact one of the purposes of these notes is to advertise Morlet's result and its significance for geometric topology. Since Morlet's proof of the lemma of disjunction was never published (however, see Millet [46] for an alternative proof) and the result itself is not widely known, we devote the first chapter to a detailed proof of the lemma of disjunction. Our proof follows in its basic strategy the proof in Morlet's notes [9], although hopefully ours is a bit more readable.

In chapters 2 and 3 we draw out some of the consequences of the lemma of disjunction for our automorphism groups. Perhaps of particular interest is the result that the first k homotopy groups of C(M), the group of pseudo-isotopies of M, depends (roughly, see Theorem B', p. 35) only on the k+3 skeleton of M. This result overlaps with the results of Cerf and Hatcher-Wagonner [21] in the case $k = 0$. Of course, their results are much sharper in the case $k = 0$ than ours since they actually compute $\pi_0 C(M)$, but their methods do not seem to generalize easily to higher k. In chapter 3 we also study some of the relationships between the automorphisms of M and those of $M \times R$ and $M \times S^1$. We show that multiplying a manifold by a torus kills off certain pathology in its automorphism groups (see Corollary 2, p. 41 for a precise statement), and yields a very useful technical device for replacing fiber spaces with manifold fibers by fiber bundles. This result is analogous to one known to and employed quite effectively by Becker, Casson, and Gottlieb in their study of generalized transfers [47].

We then go on to apply our results to the following problem: When is a map $f: V \to M$ of manifolds homotopic to a locally trivial fiber bundle? In chapter 4 we consider the case when $M = S^n$. The case $n = 2$ has been investigated by Casson [16] and we are able to generalize his results, using his techniques and our earlier results to larger n. In chapter 5 we consider the case for more general M. Here the problem is technically more formidable and our solution depends on highly non-trivial results in non-simply connected surgery theory along with some rather complicated and delicate constructions. These constructions, which are similar to unpublished constructions of Quinn [22], appear to be in themselves interesting and bear further investigation. The results of this chapter themselves can and should be further extended, for example to non simply connected M. In chapter 6 we prove prove that the automorphisms of $M \times S^1$ essentially contain the automorphisms of $M \times I$ as a direct factor. Hatcher and W.C.Hsiang have a similar result.

There are two appendices (referred to in the text as A1 and A2). The second is due to Eric Pedersen who shows how to generalize the results presented by us in the smooth and PL category to the topological category. The first appendix summarizes foundational material on n-ads, transversality, semi-simplicial sets, fiber spaces, etc. which is generally known to workers in this area but of which there is no conventient published exposition.

This work is, in general, self-contained in the sense that the results and techniques we presume the reader is familar with are available in published form and we try to give fairly complete references. The arguments and style of presentation do assume a certain familiarity with the machinery of algebraic and differential topology, and the reader should be forewarned.

The authors wish to thank Eric Pedersen who not only contributed appendix 2 but read over most of the manuscript and made several corrections and helpful suggestions.

1. Morlet's Lemma of Disjunction

We will work in both the PL (piecewise linear) and smooth categories. Almost all the results in the PL category also hold in the Top category, but the arguments require some modifications and these will be discussed in a separate appendix (A2).

Let V and W be manifolds with boundary. By an embedding $g: W \to V$ we will mean an allowable embedding $g: (W, \partial W) \to (V, \partial V)$ of manifold 2-ads (see A1); i.e. $g^{-1}(\partial V) = \partial W$ and g is transverse to the boundary and locally flat.

By a concordance $\varphi: W \times I \to V \times I$ of g we will mean an allowable embedding $\varphi: (W \times I; W \times 0, W \times 1, \partial W \times I) \to (V \times I; V \times 0, V \times 1, \partial V \times I)$ of manifold 4-ads such that $\varphi | W \times 0 \cup \partial W \times I = g \times \mathrm{id}$.

Definition 1. Let $g: W \to V$ be an embedding. The css set $E(W \times I, V \times I, g \times \mathrm{id} \bmod W \times 0 \cup \partial W \times I)$ of concordances of g will be denoted $C(W, V; g)$. This is a Kan css set (A1). I.e., an i-simplex of $C(W, V; g)$ is an allowable embedding $\varphi: \Delta_i \times W \times I \to \Delta_i \times V \times I$ of manifold i+5-ads such that

a) φ commutes with projection on Δ_i,

b) $\varphi | \Delta_i \times (W \times 0 \cup \partial W \times I) = \mathrm{id} \times g \times \mathrm{id}$.

Lemma of Disjunction. Let $g: (D^p, \partial D^p) \to (V^n, \partial V)$ and $h: (D^q, \partial D^q) \to (V^n, \partial V)$ be disjoint embeddings of discs with $n-p \geq 3$ and $n-q \geq 3$. Then

$$\pi_i(C(D^p, V; g), C(D^p, V - hD^q; g) = 0 \quad \text{for} \quad i \leq 2n-p-q-5.$$

We first observe that $C(D^p, V; g)$ and $C(D^p, V-hD^q; g)$ are connected and hence the Lemma is true for $i = 0$. This is an immediate consequence of Hudson's theorem (A1).

Notation: We will say that (i,n,p,q) is true if for any manifold V of dimension n and disjoint embeddings $g: D^p \to V$ and $h: D^q \to V$,

$$\pi_i(C(D^p, V; g), C(D^p, V - hD^q, g) = 0.$$

Sublemma 1. (i,n,p,q) implies $(i\ ,n,q,p)$.

Proof. Let $\varphi: \Delta_i \times D^q \times I \to \Delta_i \times V \times I$ represent an element of $\pi_i(C(D^q, V; h), C(D^q, V - gD^p; h)$; i.e., $\varphi_s: D^q \times I \to (V - gD^p) \times I$ for $s \in \partial\Delta_i$ and $\varphi_0 = h$. By the isotopy extension theorem (apply it first to $\varphi|\Delta_i \times D^q \times 1$) there is an automorphism $F: \Delta_i \times V \times I \to \Delta_i \times V \times I$ commuting with projection on Δ_i, and such that $\varphi_s = F_s \circ h \times id_I$ and $F_s|V \times 0 =$ identity. Then $\psi = F^{-1} \circ id_{\Delta_i} \times g \times id_I$ represents an element of $\pi_i(C(D^p, V; g), C(D^p, V - hD^q; g)$. This is isotopic mod $\partial\Delta_i$ to a simplex in $C(D^p, V - hD^q; g)$ by (i,n,p,q). Let K^t, $0 \leq t \leq 1$, be an extension of this isotopy to an ambient isotopy of $\Delta_i \times V \times I$. Then $K^1 \circ \psi$ is disjoint from $id_{\Delta_i} \times h \times id_I$.

Consider $F \circ (K^t)^{-1} \circ id_{\Delta_i} \times h \times id_I$. Then $F \circ (K^0)^{-1} \circ id_{\Delta_i} \times h \times id_I = \varphi$ since $K^0 =$ identity. Since $F \circ (K^1)^{-1} \circ K^1 \circ \psi = id_{\Delta_i} \times g \times id_I$, $F \circ (K^1)^{-1} \circ id_{\Delta_i} \times h \times id_I$ is disjoint from $id_{\Delta_i} \times g \times id_I$. This implies (i,n,q,p) since K^t is an isotopy mod $\partial\Delta_i$.

We can now reformulate the Lemma of Disjunction in an inductive manner convenient for the proof:

Sublemma 2. If $n-p \geq 3$ and $n-q \geq 3$, and if the assertion $(i-k, n, p+1, q)$ is true for $k > 0$, then (i,n,p,q) is true.

To see how this sublemma implies the Lemma of Disjunction, note that $(0,n,p,q)$ is true for $p \leq n-3$, $q \leq n-2$ by Hudson's theorem. By Sublemma 1, $(0,n,p+1,q)$ is true for $p \leq n-3$, $q \leq n-3$. Sublemma 2 then implies $(1,n,p,q)$ is true for $p \leq n-3$, $q \leq n-3$. Now assume by induction that (j,n,p,q) is true for $j < i$ and $j \leq 2n-p-q-5$, $p \leq n-3$, $q \leq n-3$ and $i > 1$. Note that $1 < i \leq 2n-p-q-5$,

$p \leq n-3$, $q \leq n-3$ requires either $p \leq n-4$ or $q \leq n-4$. If $p \leq n-4$, the induction assumption and Sublemma 2 imply (i, n, p, q) is true for $q \leq n-3$. If $q \leq n-4$, then we can use Sublemma 1 to obtain (i, n, p, q) for $p \leq n-3$. Thus we get by induction that (i, n, p, q) is true for $i \leq 2n-p-q-5$, $p \leq n-3$, $q \leq n-3$ proving the Lemma of Disjunction.

The proof of Sublemma 2 will be in four steps.

Let D_+^{p+1} be the closed upper half ball in D^{p+1}, and let S_+^p be the closed upper hemisphere in $S^p = \partial D^{p+1}$. Then $\partial D_+^{p+1} = D^p \cup S_+^p$, identified along S^{p-1}. V and $V^0 = V - h(D^q)$ will be fixed throughout the argument.

Step 1. Let $g: D^p \to V$, $h: D^q \to V$ be disjoint embeddings, $n-p \geq 3$. Suppose that g is the restriction of an allowable embedding $\bar{g}: D_+^{p+1} \to V^0$, $\bar{g}^{-1}(\partial V^0) = S_+^p$, locally flat and transverse to ∂V along S_+^p. Then $(i-1, n, p+1, q)$ implies $\pi_i(C(D^p, V; g), C(D^p, V-hD^q; g) = 0$.

Proof. We will consider g and $\bar{g}$ as inclusion $D^p \subset D_+^{p+1} \subset V^0$. We write $C(D^p) = C(D^p, V; g)$ and $C^0(D^p) = C(D^p, V^0; g)$. Let $J = [-1, 1]$ and J^r the r-fold product. Now D^p and D_+^{p+1} have trivial normal bundles $D^p \times J^{n-p} \subset V^0$ and $D_+^{p+1} \times J^{n-p-1} \subset V^0$, $D^p \times J^{n-p} \cap \partial V^0 = \partial D^p \times J^{n-p}$ and $D_+^{p+1} \times J^{n-p-1} \cap \partial V^0 = S_+^p \times J^{n-p-1}$. Further, we can assume that these coincide on $D^p \times J^{n-p}$, and that $(x, 0, \dots, 0, t_{n-p}) \in D^p \times J^{n-p}$ defines a collar neighborhood on D^p in D_+^{p+1} for $0 \leq t_{n-p} < 1$.

In this proof, for $K \subset V^0$ we will let $C(K)$ and $C^0(K)$ denote concordances of K in V and K in V^0 respectively, fixed on $K \cap \partial V^0$, even if $K \cap \partial V^0 \neq \partial K$. Thus $C(D^p \times J^{n-p})$ and $C^0(D^p \times J^{n-p})$ are the spaces of concordances which are the inclusion on $D^p \times J^{n-p} \times 0$ and $(\partial D^p) \times J^{n-p} \times I$.

Now the isotopy extension theorem implies that the restriction maps ρ:

$$\text{I}\qquad \begin{array}{ccc} C^0(D^p \times J^{n-p}) & \hookrightarrow & C(D^p \times J^{n-p}) \\ \downarrow \rho & & \downarrow \rho \\ C^0(D^p) & \hookrightarrow & C(D^p) \end{array}$$

are s. s. fibrations. The fibres are the subspaces $C^0(D^p \times J^{n-p}; D^p)$ and $C(D^p \times J^{n-p}; D^p)$ of $C^0(D^p \times J^{n-p})$ and $C(D^p \times J^{n-p})$ respectively, of concordances fixed on D^p. It is obvious that the inclusion $C^0(D^p \times J^{n-p}; D^p) \hookrightarrow C(D^p \times J^{n-p}; D^p)$ is a homotopy equivalence. Hence we have

a) $\qquad \pi_j(C(D^p \times J^{n-p}), C^0(D^p \times J^{n-p})) \cong \pi_j(C(D^p), C^0(D^p))$, all j.

Let $\overline{D}{}^{p+1}_+ = D^{p+1}_+ \cup D^p \times [-1, 0]$, where $D^p \times [-1, 0] = \{(x, 0, \ldots, 0, t_{n-1}) \in D^p \times J^{n-p} \mid -1 \leq t_{n-p} \leq 0\}$, and let $\overline{D}{}^{p+1}_+ \times J^{n-p-1} = D^{p+1}_+ \times J^{n-p-1} \cup D^p \times J^{n-p}$. Now $C(\overline{D}{}^{p+1}_+ \times J^{n-p-1})$ and $C^0(\overline{D}{}^{p+1}_+ \times J^{n-p-1})$ are the spaces of concordances fixed on $\overline{S}{}^p_+ \times J^{n-p-1}$, $\overline{S}{}^p_+ = S^p_+ \cup S^{p-1} \times [-1, 0]$. These spaces are contractible: Just deform $\overline{D}{}^{p+1} \times J^{n-p-1}$ into a collar neighborhood of $\overline{S}{}^p_+ \times J^{n-p-1}$, where we can assume (up to homotopy equivalence) that our concordances are the inclusion.

Restriction again defines fibrations

$$\text{(II)}\qquad \begin{array}{ccc} C^0(\overline{D}{}^{p+1}_+ \times J^{n-p-1}) & \hookrightarrow & C(\overline{D}{}^{p+1}_+ \times J^{n-p-1}) \\ \downarrow & & \downarrow \\ C^0(D^p \times J^{n-p}) & \hookrightarrow & C(D^p \times J^{n-p}) \end{array}$$

Let $W = V - (D^p \times \operatorname{Int} J^{n-p})$, $W^0 = V^0 - (D^p \times \operatorname{Int} J^{n-p})$. Then we can identify the fibres of II with $C(D^{p+1} \times J^{n-p-1}, W^0)$ and $C(D^{p+1} \times J^{n-p-1}, W)$ respectively, where $(D^{p+1}, \partial D^{p+1}) \subset (W^0, \partial W^0)$ is identified with $(D^{p+1}_+ - D^p \times [0, 1), D^p \times (1) \cup (S^p_+ - S^{p-1} \times [0, 1)))$. Then since the total spaces of II are contractible we have:

b) $\pi_i(C(D^p \times J^{n-p}), C^0(D^p \times J^{n-p}) \simeq \pi_{i-1}(C(D^{p+1} \times J^{n-p-1}, W), C(D^{p+1} \times J^{n-p-1}, W^0)).$

By $(i-1, n, p+1, q)$ and (a) with $p+1$ in place of p, the right side of (b) is zero. Thus the left side of (b) is zero and by (a), $\pi_i(C(D^p), C^0(D^p)) = 0$, concluding step 1.

<u>Remark.</u> In the smooth case one can get a more direct argument by using the fibrations

$$\begin{array}{ccc} C^0(D_+^{p+1}) & \hookrightarrow & C(D_+^{p+1}) \\ \downarrow & & \downarrow \\ C^0(D^p) & \hookrightarrow & C(D^p) \end{array}$$

with contractible total spaces and fibres $C^0(D_+^{p+1}; D^p)$ and $C(D_+^{p+1}; D^p)$. For in the smooth case concordances fixed on D^p can be deformed to concordances fixed on a collar neighborhood of D^p in D_+^{p+1}, since we may assume the collars are orthogonal to $D^p \times I$ and hence the concordances restricted to the collars are actually isotopies. Thus the fibres are the homotopy type of $C(D^{p+1}, W^0)$ and $C(D^{p+1}, W)$ respectively, where W^0 and W are obtained from V^0 and V by removing an open normal tube of D^p.

We put the conclusion of Step 1 in a somewhat different form. Let $g: D^p \to V^0$ be the given embedding, and suppose $\tilde{g}: D^p \times [0,1] \to V^0$, $\tilde{g}^{-1}(\partial V^0) = \partial D^p \times [0,1]$ is an (improper) embedding extending g; i.e., $\tilde{g}|D^p \times 0 = g$. For $a \in (0,1)$, denote $\tilde{g}|D^p \times (a)$ by $\tilde{g}_a: D^p \to V^0$. (Note that such extensions always exist, since $g(D^p)$ has a trivial normal bundle in V^0.)

<u>Step 1'.</u> Let $g: D^p \to V$, $h: D^q \to V$ be disjoint embeddings, $n-p \geq 3$. Let $\alpha \in \pi_i(C(D^p, V; g), C(D^p, V^0; g))$. If for some extension $\tilde{g}: D^p \times [0,1] \to V^0$ and some $a \in (0,1)$, α has a representative $\varphi: \Delta_i \to C(D^p, V; g)$ such that $\varphi(\Delta_i)$ is contained in the subspace $C(D^p, V - \tilde{g}_a D^p; g)$, then $(i-1, n, p+1, q)$ implies $\alpha = 0$.

Proof. By deleting a sufficiently small open tubular neighborhood of $\tilde{g}_a(D^p)$ from V and V^0 we get new manifolds $\overline{V}$ and $\overline{V}^0$ such that $g: D^p \to \overline{V}^0$ extends to $\overline{g}: D_+^{p+1} \to \overline{V}^0$, $\overline{g}^{-1}(\partial\overline{V}^0) = S_+^p$. (In the smooth case one needs to round the corners.) Further, our hypothesis implies that φ represents a class $\overline{\alpha} \in \pi_i(C(D^p, \overline{V}; g), C(D^p, V^0; g))$ whose image under inclusion is α. By Step 1, $\overline{\alpha}$ and hence α is zero.

Terminology: We will say that $\overline{V}$ and $\overline{V}^0$ as above are obtained from V and V^0 by "cutting along a D^p".

Unfortunately, we do not know whether any α has a representative that factors through such a cutting. However, we will show that α is a sum of a finite number of such classes. This will be done by subdividing a representative of α into small simplices each of which factors in this way.

Step 2. Let $g: D^p \to V$, $h: D^q \to V$ be disjoint embeddings, $n-p \geq 3$, $n-q \geq 3$. Then $(i-k, n, p+1, q)$, $k > 0$ and Proposition A' (below) imply (i, n, p, q).

Proposition A': Let $g_i: W_i \to V$, $g_i^{-1}(\partial V) = \partial W_i$, be embeddings, $i = 0, 1, 2, \ldots, r$, W_i compact manifolds, $\dim W_i \leq n-3$. Let φ_i be vertices of $C(W_i, V; g_i)$. Then there exist 1-simplices φ_i^t, $0 \leq t \leq 1$, of $C(W_i, V; g_i)$, $\varphi_i^0 = \varphi_i$, satisfying

a) For all t, $\varphi_i^t(W_i \times I) \cap \varphi_j^t(W_j \times I) = \emptyset$, whenever
$$\varphi_i(W_i \times I) \cap \varphi_j(W_j \times I) = \emptyset,$$

b) If $\varphi_0 = g_0 \times$ identity, then we may assume $\varphi_0^t = g_0 \times 1$, all t,

c) $\varphi_i^1(W_i \times I) = g_i(W_i) \times I$, all i.

Proof of Step 2. Fix $\alpha \in \pi_i(C(D^p), C^0(D^p))$ and let $\varphi: \Delta_i \times D^p \times I \to \Delta_i \times V \times I$ represent α. By the isotopy extension theorem there exists an i-simplex F of $C(V, V)$, $F: \Delta_i \times V \times I \to \Delta_i \times V \times I$ with $F | \Delta_i \times D^p \times I = \varphi$.

For $s \in \Delta_i$ define $F_s: V\times I \to V\times I$ by $F_s(v,t) = F(s,v,t)$. Note $F_s(D^p\times I) \subset V^0\times I$ for $s \in \partial\Delta_i$.

Using the product normal structure on $D^p \subset V^0$, we choose $D^p\times[0,2] \subset V^0$, with $D^p\times(0) = D^p$ and $F_s(D^p\times[0,2]\times I) \subset V^0\times I$ for $s \in \partial\Delta_i$. Choose 3^i numbers $1 = a_0 < a_1 < \ldots < a_{3^i-1} < 2$. Let Δ_i' be a sufficiently fine subdivision of Δ_i so that:

a) $F_{s_1}(D^p\times[1/2,2]\times I)$ and $F_{s_2}(D^p\times I) = \varphi_{s_2}(D^p\times I)$ are disjoint for s_1, s_2 in the same simplex of Δ_i'.

b) If x is any vertex of Δ_i' and $s_1, s_2 \in St(x)$, then

$$F_{s_1}(D^p\times a_j\times I)\cap F_{s_2}(D^p\times a_k\times I) = \emptyset \quad \text{for } j\neq k.$$

c) If x is any vertex in $\partial\Delta_i'$ and $s \in St(x)$, then

$$F_s(D^p\times[0,2]\times I) \subset V^0\times I.$$

(Here St means the closed star in the complex Δ_i'.)

Choose a function $n(\ )$ from the vertices of Δ_i' to $\{0,1,\ldots,3^i-1\}$ such that $n(y)\neq n(z)$ if $y,z \in St(x)$ and $y\neq z$. This is possible by triangulating by planes parallel to the faces and indexing the vertices by $i+1$-tuples of mod 3 numbers indicating the number of parallel planes between the point and a face of Δ_i. Write $a(x) = a_{n(x)}$.

We assume we are given initially, for each vertex $x \in \Delta_i'$, an isotopy $F_x^t: D^p\times a(x)\times I \to V\times I$, $0\leq t\leq 1$, with $F_x^0 = F_x|D^p\times a(x)\times I$ and $F_x^1(D^p\times a(x)\times I) = Id|D^p\times a(x)\times I$ such that F_x^t is a 1-simplex of $C(D^p\times a(x), V)$ and:

1) $F_x^t(D^p\times a(x)\times I)\cap F_y^t(D^p\times a(y)\times I) = \emptyset$ whenever

$$F_x(D^p\times a(x)\times I)\cap F_y(D^p\times a(y)\times I) = \emptyset.$$

2) $F_x^t(D^p\times a(x)\times I)\cap V^0\times I$ for $x \in St(y)$, $y \in \partial\Delta_i'$.

Denoting by $S^m(\Delta_i)$ the m-skeleton of Δ_i' and letting $\partial S^m(\Delta_i) = S^m(\Delta_i) \cap \partial\Delta_i'$, we will construct inductively a sequence of isotopies φ_m^t, $0 \le t \le 1$, $m = 0,\ldots, i$, $\varphi_m^t : S^m(\Delta_i) \to C(D^p, V)$, $\varphi_m^t(\partial S^m(\Delta_i)) \subset C(D^p, V)$, $\varphi_m^0 = \varphi | S^m(\Delta_i)$, $\varphi_m^1(S^m(\Delta_i) \times D^p \times I) \subset V^0 \times I$, and $\varphi_m^t = \varphi_m^1$ for $\frac{m+1}{i+1} \le t \le 1$. Further, the φ_m^t will satisfy

3) For any simplex $\sigma \in S^m(\Delta_i)$ and any vertex $x \in St(\sigma)$,

$$\varphi_m^t(\sigma \times D^p \times I) \cap F_x^t(D^p \times a(x) \times I) = \emptyset .$$

The existence of the isotopies F_x^t and φ_0^t satisfying 1), 2), and 3) follows from Proposition A' and conditions a), b), c) above. We assume inductively that φ_{m-1}^t is defined and satisfies 3) with respect to the now given F_x^t. (Note $F_x^t = F_x^1$, $\frac{1}{i+1} \le t \le 1$.)

Let $\sigma \in S^m(\Delta_i)$ be an m-simplex and let $x_1, \ldots, x_d$ be the vertices in $St(\sigma)$. Let $K = \sigma \cup \bigcup_{i=1}^{d} x_i \times \sigma$ be d+1 disjoint m-simplices, and let $K_0 = \partial\sigma \cup \bigcup_{i=1}^{d} x_i \times \sigma \subset K$. Define a map $\lambda : K \times D^p \times I \to V \times I$, $\lambda | \sigma \times D^p \times I = \varphi | \sigma \times D^p \times I$, and $\lambda | x_i \times \sigma \times D^p \times I$ defined by $\lambda(x_i, s, b, r) = F_{x_i}(b, a(x_i), r)$. Then λ is a family of embeddings $D^p \times I \to V \times I$ parameterized by K. Note that λ may also be viewed as d+1 disjoint embeddings of $\sigma \times D^p \times I \to \sigma \times V \times I$, commuting with projection on σ. Then the $F_{x_i}^t$ and φ_{m-1}^t define an isotopy of $\overline{\lambda} = \lambda | K_0$, $\overline{\lambda}_t$, $0 \le t \le \frac{m}{i+1}$, through disjoint embeddings, by 1), 2), and 3) for m-1. Further if $\sigma \in \partial S^m(\Delta_i')$, the isotopies are all in V^0.

By the isotopy extension theorem applied to $\sigma \times V \times I$ (or $\sigma \times V^0 \times I$ if $\sigma \in \partial S^m(\Delta_i)$), there exists an extension of $\overline{\lambda}_t$ to $\lambda_t : K \times D^p \times I \to V \times I$; an isotopy through disjoint embeddings which are in V^0 if $\sigma \in \partial S^m(\Delta_i)$. In particular, $\lambda_1(\sigma \times D^p \times I) \subset V - \bigcup_{i=1}^{d} D^p \times a(x_i)$, and $\lambda_1(\partial\sigma \times D^p \times I) \subset V^0 - \bigcup_{i=1}^{d} D^p \times a(x_i)$.

Let W, W^0 be the manifolds obtained from V, V^0 by removing $D^p \times a(x_i)$ for all but the smallest $a(x_i)$, and let $\overline{W}, \overline{W}^0$ be the mainfolds obtained from W, W^0 by cutting along the remaining $D^p \times a(x_i)$. Then $\lambda_1 | \sigma \times D^p \times I$ represents an element of $\pi_m(C(D^p, W), C(D^p, W^0))$ which satisfies the hypothesis of Step 1' and hence is trivial. In fact, it represents the trivial element in $\pi_m(C(D^p, \overline{W}), C(D^p, \overline{W}^0))$, and hence $\lambda_1 | \sigma \times D^p \times I$ is isotopic in $V - \bigcup_{i=1}^{d} D^p \times a(x_i)$ to λ_2 in $V^0 - \bigcup_{i=1}^{d} D^p \times a(x_i)$, and hence by an isotopy satisfying 3) and fixed on $\partial\sigma$.

Define $\varphi_m^t | \sigma \times D^p \times I$ by the isotopy

$$\lambda \text{ to } \lambda_1, \quad \text{for } 0 \leq t \leq \frac{m}{i+1},$$

$$\lambda_1 \text{ to } \lambda_2, \quad \text{for } \frac{m}{i+1} \leq t \leq \frac{m+1}{i+1}.$$

Then $\varphi_m^t | \partial\sigma \times D^p \times I = \varphi_{m-1}^t | \partial\sigma \times D^p \times I$, so the isotopies fit together to give an isotopy $\varphi_m^t : S^m(\Delta_i) \times D^p \times I \to V \times I$, satisfying 3).

Step 3. Proposition A' is a consequence of Proposition A (below).

Proposition A. Let V be an n-manifold, and for $i = 0, 1, \ldots, r$ let W_i be compact manifolds with $\dim W_i \leq n-3$. Let

$$\varphi_i : (W_i \times I; W_i \times 0, W_i \times 1) \to (V \times I; V \times 0, V \times 1)$$

be embeddings in general position such that $\varphi_i | \partial W_i \times I = g_i | \partial W_i \times \text{id}$, where $g_i : W_i \to V$ is an embedding defined by $\varphi_i(x, 0) = (g_i(x), 0)$.

Then there exist isotopies $\varphi_i^t, \varphi_i^0 = \varphi_i$, fixed on $\partial W_i \times I$ such that

a) For all t, $\varphi_{i_1}^t(W_{i_1} \times I) \cap \ldots \cap \varphi_{i_s}^t(W_{i_s} \times I) = \emptyset$ whenever

$$\varphi_{i_1}(W_{i_1} \times I) \cap \ldots \cap \varphi_{i_s}(W_{i_s} \times I) = \emptyset .$$

b) The φ_i^1 are in general position.

c) There exists a concordance $H : V \times I \to V \times I$ such that for all i,

$$H(g_i^1(W_i) \times I) = \varphi_i^1(W_i \times I).$$

d) If $\varphi_0 = g_0 \times \text{id}$, then we may assume $\varphi_0^t = g_0 \times \text{id}$, all t, and $H | g_0(W_0) \times I = \text{id}$.

<u>Proof that Proposition A implies Proposition A'.</u> Let $\varphi_i^0 = \varphi_i$ satisfy the hypothesis of Proposition A'.

1. First for $i > 0$, deform φ_i^0 to φ_i^1 via transversality arguments so that the hypotheses of Proposition A are satisfied. Do this carefully so that $\varphi_i^t | \partial W_i \times I = g_i^t \times \mathrm{id}$.

In the PL category we must use block transversality (A2), since we are dealing with manifolds with boundary.

2. Apply Proposition A to deform φ_i^1 to φ_i^2 satisfying a), b), c), and d) of Proposition A. In particular, there exists a concordance $H: V \times I \to V \times I$ such that $H(g_i^2(W_i) \times I) = \varphi_i^2(W_i \times I)$. (And $H | g_0(W_0) \times I = \mathrm{id}$ if $\varphi_0 = g_0 \times \mathrm{id}$.)

3. Since $\bigcup g_i^2(W_i)$ is a codimension 3, by Hudson's Theorem, there is an ambient isotopy $R^t: V \times I \to V \times I$, fixed on $\partial V \times I \cup V \times 0$ (and on $g_0(W_0) \times I$) such that $R^0 = \mathrm{id}$ and $R^1 \circ H \circ g_i^2 \times \mathrm{id} = g_i^2 \times \mathrm{id}$. Define an isotopy of φ_i^2 to φ_i^3 by $\varphi_i^t = R^{t-2} \varphi_i^2$. Then $\varphi_i^3(W_i \times I) = g_i^2(W_i) \times I$.

4. Let $\varphi_i^t = (g_i^{5-t} \times \mathrm{id})\ (g_i^2 \times \mathrm{id})^{-1} \circ \varphi_i^3$, $3 \leq t \leq 5$. Then $\varphi_i^5(W_i \times I) = g_i(W) \times I$. Thus φ_i^t, $0 \leq t \leq 5$ would be the required isotopy if it were constant on $\partial W_i \times I \cup W_i \times 0$.

5. In any case, φ_i^t is of the form $g_i^t \times \mathrm{id}$ on $\partial W_i \times I \cup W_i \times 0$, where $g_i^t = g_i^{5-t}$ for $0 \leq t \leq 2$ and constant for $2 \leq t \leq 3$. Note g_i^t can be deformed to the constant isotopy mod endpoints.

Now we can assume by a small initial deformation, that $\varphi_i = g_i \times \mathrm{id}$ on $W_i \times [0, \epsilon]$, some $\epsilon > 0$. Then apply 1-4 to $\varphi_i | W_i \times [\epsilon, 1] \to V \times [\epsilon, 1]$, and use the deformation of g_i^t to the constant isotopy on the interval $[0, \epsilon]$ to obtain φ_i^t with φ_i^t constant on $W_i \times 0$, and satisfying a) of Proposition A'. Note that $\varphi_i^t | \partial W_i \times I$ retains the above form.

Apply the same argument to an ϵ-neighborhood of $\partial W_i \times I$ in $W_i \times I$. We then obtain φ_i^t satisfying a), b), and c) of Proposition A'.

Step 4. Proof of Proposition A.

The idea of the proof is to modify the intersections $\varphi_{i_1}(W_{i_1} \times I) \cap \ldots \cap \varphi_{i_s}(W_{i_s} \times I)$ until they are of the form

$$\varphi'_{i_1}(W_{i_1} \times I) \cap \ldots \cap \varphi'_{i_s}(W_{i_s} \times I) \cong (g'_{i_1}(W_{i_1}) \cap \ldots \cap g'_{i_s}(W_{i_s})) \times I.$$

One uses this product structure to build up a product structure on $\bigcup \varphi'_i(W_i \times I)$, starting with the smallest dimensional intersections. This product structure is extended to a new product structure on $V \times I$, defining the concordance H. The technique for modifying intersections is the Whitney trick.

The argument involves a long complicated induction. Therefore, the reader is urged to look at the case where there are only two manifolds, W_0 and W_1, in order to understand the essentials of the proof.

Proof of Proposition. Set $L_i = \varphi_i(W_i \times I)$ and for $i_1 < i_2 < \ldots < i_p$, $L_{i_1 i_2 \ldots i_p} = L_{i_1} \cap L_{i_2} \cap \ldots \cap L_{i_p}$. Order the indices lexicographically and write $X_1, X_2, \ldots, X_s$ for the $L_{i_1 i_2 \ldots i_p}$ in this order.

For any subset A of $V \times I$, set $A^0 = A \cap (V \times 0)$, $A^1 = A \cap (V \times 1)$, $dA = A \cap (\partial V \times I)$, $dA^0 = A^0 \cap (\partial V \times 0)$, $dA^1 = A^1 \cap (\partial V \times 1)$. Then (X_j, X_j^0, X_j^1) is a cobordism such that (dX_j, dX_j^0, dX_j^1) is the product cobordism $(dX_j^0 \times I, dX_j^0 \times 0, dX_j^0 \times 1)$, compatible with the product structure $\partial V \times I$.

If $Y \subset V \times I$ is any such cobordism, i.e., $dY = dY^0 \times I$ compatibly, one says that $\tau(Y) \geq \alpha$ if the following holds:

> Y may be constructed from a regular neighborhood of Y^0 by adding handles (not meeting dY^0) of dim i, $\alpha \leq i \leq \dim Y - \alpha$.

In particular, if $\tau(Y) > \frac{1}{2} \dim Y$, Y is isomorphic to $Y_0 \times I$.

Lemma 1.1. Let $Y \subset V \times I$ be a cobordism, Y connected and of dim n. Let B be a subcomplex of Y with $dB = dB^0 \times I$, B of codim ≥ 3 in Y and B^0 of codim ≥ 3 in Y^0. Let $\tilde{Y}$ be the universal cover of Y and $\tilde{Y}^0, \tilde{Y}^1, \tilde{B}, \tilde{B}^0, \tilde{B}^1$ the preimage of Y^0, Y^1, B^0, B^1 in $\tilde{Y}$. If (cohomology with compact supports)

1. $H^{n-1-k}(\tilde{B}, \tilde{B}^1) = 0$ for $k \leq \alpha$, and
2. $\pi_k(Y, Y^0) = 0$ for $k \leq \alpha - 1$;

then $j_*: \pi_i(Y - B, Y^0 - B) \to \pi_i(Y, Y^0)$ is an isomorphism for $i \leq \alpha$, $\alpha \neq 2$. If $\alpha = 2$, j_* is an isomorphism for $i < 2$ and an epimorphism for $i = 2$, it will be an isomorphism for $i = 2$ if we further assume $\pi_2(Y, Y^0) = 0$.

Proof. Since B has codim ≥ 3, j_* is an epimorphism for $i \leq 2$ and a monomorphism if $i \leq 1$, by transversality. Also $\pi_i(Y, Y - B) = \pi_i(Y^0, Y^0 - B) = 0$ for $i \leq 2$. If $\alpha > 2$ or if $\alpha = 2$ and $\pi_2(Y, Y^0) = 0$, $\pi_1(Y) = \pi_1(Y^0) = \pi_1(Y - B) = \pi_1(Y^0 - B^0)$. Let $T(T^0)$ be a regular neighborhood of $B(B^0)$ in $Y(Y^0)$, and $\tilde{T}(\tilde{T}^0)$ the preimage in $\tilde{Y}$. Then $\tilde{Y}, \tilde{Y}^0, \tilde{Y} - \tilde{T}$ and $Y^0 - \tilde{T}^0$ are 1-connected. Therefore, it suffices to show that $H_k(\tilde{Y} - \tilde{T}, \tilde{Y}^0 - \tilde{T}^0) = 0$ if $k \leq \alpha - 1$ and $H_\alpha(\tilde{Y} - \tilde{T}, \tilde{Y}^0 - \tilde{T}^0) \to H_\alpha(\tilde{Y}, \tilde{Y}^0)$ is an isomorphism. By Poincaré duality, since $\partial Y = Y^0 \cup Y^1 \cup dY^0 \times I$, it suffices to show $H^{n-k}(\tilde{Y}, \tilde{T} \cup \tilde{Y}^1) = 0$ if $k \leq \alpha - 1$, and that $H^{n-\alpha}(\tilde{Y}, \tilde{T} \cup \tilde{Y}^1) \to H^{n-\alpha}(\tilde{Y}, \tilde{Y}^1)$ is an isomorphism. Consider the exact sequence of the triple $(\tilde{Y}, \tilde{T} \cup \tilde{Y}^1, \tilde{Y}^1)$:

$$H^{n-k}(\tilde{T} \cup \tilde{Y}^1, \tilde{Y}^1) \leftarrow H^{n-k}(\tilde{Y}, \tilde{Y}^1) \leftarrow H^{n-k}(\tilde{Y}, \tilde{T} \cup \tilde{Y}^1) \leftarrow H^{n-k-1}(\tilde{T} \cup \tilde{Y}^1, \tilde{Y}^1).$$

But $H^{n-k-1}(\tilde{Y}^1 \cup \tilde{T}, \tilde{Y}^1) = H^{n-k-1}(\tilde{T}, \tilde{T} \cap \tilde{Y}^1) = H^{n-k-1}(\tilde{T}, \tilde{T}^1) = 0$ if $k \leq \alpha$. Hence the result also holds for $\alpha > 2$.

Lemma 1.2. Let $M_1, \dots, M_s$ be codimension 3 cobordisms, $dM_i = dM_i^0 \times I$, in Y^n in general position and assume for any $(j_1, \dots, j_k) \subset (1, 2, \dots, s)$

$$\tau(M_{j_1} \cap \dots \cap M_{j_k}) \geq \alpha - 1.$$

Let $\widetilde{Y}$ be the universal cover of Y and $\widetilde{M}_i$ the preimage of M_i in $\widetilde{Y}$. Then

$$H^k(\widetilde{M}_1 \cup \dots \cup \widetilde{M}_s; \widetilde{M}_1^1 \cup \dots \cup \widetilde{M}_s^1) = 0 \quad \text{for } k > \max \dim(M_i) - \alpha + 1$$
$$\geq n - \alpha - 2 \ .$$

Proof. M_i is obtained from M_i^1 by adding handles of dim j, $\alpha - 1 \leq j \leq \dim M_i - \alpha + 1$. Hence $\widetilde{M}_i$ is obtained from $\widetilde{M}_i^1$ by adding handles in this dimension range, so $H^k(\widetilde{M}_i, \widetilde{M}_i^1) = 0$ for $k > \dim M_i - \alpha + 1$. Assume the result for $B_{r-1} = M_1 \cup \dots \cup M_{r-1}$. Then the Mayer-Vietoris sequence:

$$\to H^{k-1}(B_{r-1} \cap M_r, B_{r-1}^1 \cap M_r^1) \to H^k(B_r, B_r^1) \to H^k(B_{r-1}, B_{r-1}^1) \oplus H^k(M_r, M_r^1) \to$$

gives the result for B_r. In fact, $B_{r-1} \cap M_r = \bigcup_{i=1}^{r-1} (M_i \cap M_r)$ and so applying the result for $r-1$ terms to $M_i \cap M_r \subset M_r$, the term at the left is zero for $k - 1 > \max \dim M_i \cap M_r - \alpha + 1 \geq \max \dim M_i - \alpha - 2$.

Lemma 1.3. If $\tau(X_j) \geq \alpha$ for $j < m$

$$\geq \alpha - 1 \text{ for } j \geq m$$

then there exist isotopies φ_i^t satisfying a) and b) of Proposition A such that $X_j' = X_j$ for $j < m$ and $\tau(X_m') \geq \alpha$. Further, if $\varphi_0 = g_0 \times \mathrm{id}$, then we may assume $\varphi_0^t = g_0 \times \mathrm{id}$, all t (where X_j' etc. corresponds to the situation when $t = 1$).

Proof of Lemma 1.3. If $X_m = \emptyset$ or if X_m is an L_r, then there is nothing to prove. Hence we may assume $X_m = X_n \cap L_r$, where $X_n = L_{i_1} \cap \dots \cap L_{i_q}$, $i_p < r$ for all $p \leq q$. Let Z be the union of all X_j which do not contain X_m.

We have $\tau(X_m) \geq \alpha - 1$. We may assume $\alpha - 1 \leq \frac{1}{2} \dim X_m$. Let $\mu_s : D^{\alpha-1} \to X_m$ be a finite family of disjoint embeddings of

$(D^{\alpha-1}, S^{\alpha-2}) \to (X_m, X_m^0)$ or (X_m, X_m^1) as cores of all α-1 handles of X_m mod X_m^0 or X_m^1. Since $\tau(X_j) \geq \alpha-1$, all j, we have $\pi_{\alpha-1}(X_m - Z, X_m^0 - Z^0) \to \pi_{\alpha-1}(X_m, X_m^0)$ is an epimorphism by Lemmas 1.1, 1.2 (similarly for X_m^1). Hence if $2(\alpha-1) \leq \dim X_m$ and $\dim X_m \geq 5$ we may assume by Hudson's embedding theorem (A1) that Image μ_s is in $X_m - Z$. But if $\dim X_m \leq 4$, $\dim Z \cap X_m < 1/2 \dim X_m$, and so we may still assume Im μ_s is in $X_m - Z$ by general position.

Consider the exact sequences:

$$\to \pi_\alpha(X_n - Z; X_n^0 - Z^0, X_m - Z) \to \pi_{\alpha-1}(X_m - Z, X_m^0 - Z^0) \to \pi_{\alpha-1}(X_n - Z, X_n^0 - Z^0) \to$$

$$\to \pi_\alpha(L_r - Z; L_r^0 - Z^0, X_m - Z) \to \pi_{\alpha-1}(X_m - Z, X_m^0 - Z^0) \to \pi_{\alpha-1}(L_r - Z, L_r^0 - Z^0) \to \ .$$

By Lemma 1.1 and the fact that $\pi_{\alpha-1}(L_r, L_r^0) \simeq \pi_{\alpha-1}(X_n, X_n^0) = 0$ (similarly for X_m^1), the right-hand groups are zero. Since $2\alpha < \dim L_r$ or $\dim X_n$, there exist disjoint embeddings $f_s : D^\alpha \to L_r - Z$, $f_s(S_-^{\alpha-1}) \subset L_r^0$ or L_r^1, $f_s | S_+^{\alpha-1} = \mu_s$, $\operatorname{Im} f_s \cap X_m = \operatorname{Im} \mu_s$; and $g_s : D^\alpha \to X_n - Z$, $g_s(S_-^{\alpha-1}) \subset X_n^0$ or X_n^1, $g_s | S_+^{\alpha-1} = \mu_s$, $\operatorname{Im} g_s \cap X_m = \operatorname{Im} \mu_s$.

From the exact sequence

$$\to \pi_{\alpha+1}(V \times I - Z, L_r \cup X_n - Z, V \times 0 - Z^0) \to \pi_\alpha(L_r \cup X_n - Z, L_r^0 \cup X_n^0 - Z^0)$$

$$\to \pi_\alpha(V \times I - Z, V \times 0 - Z^0) \to$$

and similar arguments, there exist disjoint embeddings $F_s : D_+^{\alpha+1} \to V \times I - Z$ such that $F_s(S_+^\alpha) \subset V \times 0$ or $V \times 1$, $F_s | D_+^\alpha = f_s$, $F_s | D_-^\alpha = g_s$.

Now deform L_r (i.e. φ_r) by pushing along Im F_s to remove a regular neighborhood of Im μ_s from X_m. Then $\tau X_m' \geq \alpha$. If $j < m$, and j contains the index r, then j contains an indice not in m, and $X_j \subset Z$. Hence $X_j' = X_j$. But also if $j < m$ and j does not contain the index r, $X_j' = X_j$.

Our deformation of L_r leaves it transverse to X_n and all the X_j's that do not contain X_m. For the rest, i.e., $L_{i_1}, \ldots, L_{i_q}$ and their intersections, it may no longer be transverse. However, L'_r is transverse to the $X_j \supset X_m$ near X_n, and hence may be deformed slightly to be transverse to the $X_j \supset X_m$ without changing its intersection with X_n or the other X_j's.

Corollary 1.4. There exist isotopies φ_i^t satisfying a) and b) of Proposition A such that $\tau(X_j') > \frac{1}{2} \dim X_j'$ for all j. Further, if $\varphi_0 = g_0 \times \mathrm{id}$, then we may assume $\varphi_0^t = g_0 \times \mathrm{id}$, all t.

Proof. First, $\tau(X_1) \geq \alpha$ all α, since $X_1 = L_0$. Assume $\tau(X_j) \geq \alpha_j$, and $\alpha_j \geq \alpha_{j+1}$. By induction on the lexicographic order of $(\alpha_1, \alpha_2, \ldots, \alpha_s)$ and maintaining the condition $\alpha_j \geq \alpha_{j+1}$, we can apply (1.3) to raise the order by one.

Lemma 1.5. Assume $X_1, \ldots, X_s \subset V \times I$ are cobordisms as in Proposition A; i.e. in general position and closed under intersection. If $\tau(X_j) \geq \frac{1}{2} \dim X_j$, all j; then $X_j \cong X_j^0 \times I$ by compatible isomorphisms.

Proof. The condition $\tau(X_j) > \frac{1}{2} \dim X_j$ implies each X_j has a product structure, but these are not necessarily compatible. We modify these product structures by induction on $\dim X_j$: The lowest dimensional X_j's don't intersect and hence their union is a product. Assume that $X_{i_1}, \ldots, X_{i_r}$ are the cobordisms of dimension less than k and they are compatibly isomorphic to a product. Let X_j be a cobordism of dimension k. $X_j \cap (X_{i_1} \cup \ldots \cup X_{i_r})$ is a product since it is the union of X_i's of lower dimension, in fact codimension at least 3. By Hudson's Theorem -- at least in the PL case -- there is an ambient isotopy of X_j giving a new product structure on X_j, extending the product structure on this subcomplex. Since two k-dimensional cobordisms meet in a lower dimensional cobordism, this gives compatible product structures on all cobordisms of dimension $\leq k$, completing our induction step.

In the smooth case, first choose a Riemannian metric so that all the X_i's meet orthogonally. Then we may apply Hudson to the smallest intersections X_i in X_j to obtain a product structure on X_j which extends that on the normal tubes of these X_i's in X_j. Deleting the interiors of these normal tubes, we may proceed by induction to apply Hudson's theorem to obtain a compatible product structure on X_j by an ambient isotopy of X_j. Thus the result holds in the smooth case as well.

By applying the same arguments to the union of the X_j's in $V \times I$ we get:

Corollary 1.6. With $X_1, \ldots, X_s$ as in 1.5, there exists a concordance $H: V \times I \to V \times I$ such that $H(X_j^0 \times I) \cong X_j$, all j. Further, with respect to a given product structure on X_1, $\varphi: X_1^0 \times I \to X_1$, $H|X_1^0 \times I$ is isotopic to φ.

Proof of Proposition A. Applying 1.6 to the result of 1.4, we get isotopies φ_i^t satisfying a), b), and c). Further, $\varphi_0^t = g_0 \times \mathrm{id}$, all t. But $H|g_0(W_0) \times I$ is only isotopic to the identity. Let R^t be an ambient isotopy of $V \times I$ fixed on $V \times 0 \cup \partial V \times I$ extending this isotopy; i.e. R^0 = identity, $R^1|g_0(W_0) \times I = H|g_0(W_0) \times I$. Then $(R^t)^{-1} \circ \varphi_i^t$, $i > 0$, $\varphi_0^t = g_0 \times \mathrm{id}$, are new isotopies satisfying a) and b). But $(R^1)^{-1} \circ H(g_i(W_i) \times I) = (R^1)^{-1} \circ \varphi_i^1(W_i \times I)$, $i > 0$, and $(R^1)^{-1} \circ H|g_0(W_0) \times I$ = identity.

We give another version of Step 2, leading to some improvements in the Lemma of Disjunction, especially in the 1-connected case.

Step 2'. Let $g: D^p \to V$, $h: D^q \to V$ be disjoint embeddings, $n-p \geq 3$. Let $\alpha \in \pi_i(C(D^p, V; g), C(D^p, V-hD^q; g))$ be in the image of $\pi_i(C(D^p, V; g))$. Then $(i-k, n, p+1, q)$, $k > 0$, and Proposition A' implies $\alpha = 0$.

Proof. Take a representative $\varphi: \Delta_i \times D^p \times I \to \Delta_i \times V \times I$ of α such that $\varphi|\partial\Delta_i \times D^p \times I = \mathrm{id} \times g \times \mathrm{id}$. With F satisfying a), b), and c) as in Step 2, assume initially we are given isotopies F_x^t, for each vertex $x \in \Delta_i'$ not in $\partial\Delta_i'$ satisfying 1). Construct inductively isotopies $\varphi_m^t: S^m(\Delta_i) \to C(D^p, V)$ as before with the further condition that $\varphi_m^t | \partial S^m(\Delta_i) = \mathrm{id} \times g \times \mathrm{id}$, and satisfying 3) for $x \notin \partial\Delta_i'$. Applying Proposition A' with $\varphi_0 = g \times \mathrm{id}$ (instead of $h \times \mathrm{id}$) to obtain F_x^t and φ_0^t satisfying 3). The induction proceeds as in Step 2, except that for $\sigma \in \partial S^m(\Delta_i)$ instead of requiring $\overline{\lambda}_t$ to be in V^0, we define $\overline{\lambda}_t$ as the constant map $\mathrm{id} \times g \times \mathrm{id}$ on $\sigma \times D^p \times I$.

Remark. If $i = 1$, the condition on α is always fulfilled since $\pi_0(C(D^p, V\text{-}hD^q; g) = 0$ by Hudson. Hence $(1, n, p, q)$ holds for $p < n-3$, any q.

Addendum to Lemma of Disjunction: If V^n is simply connected, $n \geq 5$, then the result holds for $n-p \geq 3$, $n-q \geq 3$ and $i \leq 2n-p-q-4$.

Our starting point is:

Proposition. Let $g: (D^{n-2}, S^{n-3}) \to (V, \partial V)$ be an embedding, $\dim V = n \geq 5$. If $\varphi: D^{n-2} \times I \to V \times I$ is a concordance of g which is trivial in the boundary and satisfies

$$(*) \quad \pi_1(V - \varphi_0(D^{n-2})) = \pi_1(V - \varphi_1(D^{n-2})) = \pi_1(V \times I - \varphi(D^{n-2} \times I) = 0,$$

then there is an ambient isotopy F of $V \times I$ fixed on $\partial V \times I \cup V \times 0$, such that $F_0 =$ identity and $F_1 \circ \varphi = g \times \mathrm{id}$.

Proof. This follows from Rourke [11]. Alternatively, since φ extends to the normal tube, i.e. $\tilde{\varphi}: D^{n-2} \times D^2 \times I \to V \times I$, and since $V \times I - \tilde{\varphi}(D^{n-2} \times D^2 \times I)$ is an h-cobordism by $(*)$, we may extend $\tilde{\varphi}$ to a concordance $\tilde{\varphi}: V \times I \to V \times I$. Then we can apply Cerf [4] in the smooth case and [3] or Hatcher (unpublished) in the PL-case to deform $\tilde{\varphi}$ to the identity and hence φ to $g \times \mathrm{id}$.

Now suppose $g: (D^{n-3}, S^{n-4}) \to (V, \partial V)$ is a trivial embedding; i.e g extends to $\overline{g}: D_+^{n-2} \to V$ and hence to $\tilde{g}: D_+^{n-2} \times J^2 \to V$. As in Step 1 of the proof of Sublemma 2, let $W = V - \tilde{g}(D^{n-3} \times \mathrm{Int}\, J^3)$, and consider the resulting embedding $\tilde{g}: (D^{n-2}, S^{n-3}) \to (W, \partial W)$. Then $\pi_1(W - \tilde{g}D^{n-2}) = \pi_1(V - gD_+^{n-2}) = 0$.

Similarly, if $\varphi: D^{n-3} \times I \to V \times I$ is a concordance of g which extends to $\tilde{\varphi}: D_+^{n-2} \times J^2 \times I \to V \times I$, $\tilde{\varphi}|D^{n-3} \times J^3 \times I = \tilde{g} \times id$, the resulting concordance $\tilde{\varphi}: D^{n-2} \times I \to W \times I$ satisfies $(*)$.

Consequently, $(1, n, p, q)$, $p \leq n-3$, $q \leq n-2$ follows, if V is 1-connected. An induction argument similar to that following Sublemma 2 proves the Addendum.

__Theorem 1.7.__ Let $(V, \partial V)$ be a compact n-manifold, $n \geq 5$, with non-empty boundary which is k-connected, $k \geq 2$. Let $f: (D^p, S^{p-1}) \to (V, \partial V)$ be an embedding, $p \leq n-3$, and let $\bar{f}: (D^p \times D^{n-p}, S^{p-1} \times D^{n-p}) \to (V, \partial V)$ be an extension to a closed tubular neighborhood. Then if $N^n = \bar{f}(D^p \times D^{n-p})$, $\pi_i(C(D^p, D^n; i) = \pi_i(C(D^p, N; f) \to \pi_i(C(D^p, V; f)$ is an isomorphism for $i \leq n+k-p-4$ and an epimorphism for $i \leq n+k-p-3$; provided $k \leq n-4$ if $n = 5$ or ∂V not 1-connected.

__Proof.__ $P = V - \mathring{N}$ may be viewed as a cobordism rel boundary with $\partial_- P = D^p \times S^{n-p-1}$. Since $H_i(P, \partial_- P) = H_i(V, N) = H_i(V)$, $i > 0$, and V is k-connected; V may be built up from N by adding handles of dimension $> k$ to the interior of $\partial N = \partial_- P$. This requires $k \leq n-4$ if $\partial_+ P$ is not 1-connected. But $\partial_+ P \cong \partial V - S^{p-1} \times \mathring{D}^{n-p}$ and $\pi_i(\partial_+ P) \cong \pi_i(\partial V)$ for $i = 0, 1$, since $p \leq n-3$.

Now this means that $P = V - \mathring{N}$ is built up from $\partial_+ P$ by adding handles $h_1, \ldots, h_r$ of dimension $\leq n-k-1$ to the interior of $\partial_+ P$. Let N_j be the closure of $V - \partial_+ P \times I \cup h_1 \cup \ldots \cup h_j$. Thus $N_0 = V$ and $N_r = N$. By the Addendum to the Lemma of Disjunction $\pi_i(C(D^p, N_{j+1})) \to \pi_i(C(D^p, N_j))$ is an isomorphism for $i \leq n+k-p-4$ and an epimorphism for $i \leq n+k-p-3$, $k \geq 2$. (Note that N_j is at least k-connected since it is obtained from N by adding handles of dimension $> k$.) The result follows by induction.

Remarks. 1. If we know V is built up from N by adding handles of dimension $> k$ to the interior of $\partial N = \partial_- P$, then the result holds for $n \geq 5$ regardless of the connectivity of ∂V. In particular, for $V = S^{n-1} \times D^1$ we have

$$\pi_i(C(D^1, D^n; i) \to \pi_i(C(D^1, S^{n-1} \times D^1)) \text{ is an isomorphism}$$

for $i \leq 2n-7$ and an epimorphism for $i \leq 2n-6$.

2. The theorem also holds for $k = 0, 1$, as will be shown in Chapter 2.

2. Lemma of Disjunction -- 2nd form

Let $g: (W, \partial W) \to (V, \partial V)$ be an embedding. Let $E(W, V; g)$ and $\widetilde{E}(W, V; g)$ be the Δ-sets of isotopies and pseudoisotopies of W in V fixed on ∂W (i.e. $E(W, V; g \bmod \partial W)$, see A1). Let $\widetilde{C}(W, V; g) = \widetilde{E}(W \times I, V \times I, g \times id \bmod W \times 0 \cup \partial W \times I)$. These are Kan Δ-sets (A1); for W compact the restriction map $C(W, V; g) \to E(W, V; g)$ is a Kan fibration and the restriction map $\widetilde{C}(W, V; g) \to \widetilde{E}(W, V; g)$ is Kan if $\dim V - \dim W \geq 3$ (A1). The fibres are $E(W \times I, V \times I; g \times id)$ and $\widetilde{E}(W \times I, V \times I; g \times id)$ respectively.

If V is also compact we have the Kan fibrations $A(V) \to E(W, V)$ and $\widetilde{A}(V) \to \widetilde{E}(W, V)$, when $A(V)$ and $\widetilde{A}(V)$ are the automorphisms fixed on ∂V (i.e. $A(V \bmod \partial V)$, $\widetilde{A}(V, \bmod \partial V)$, see A1). The fibres are $A(V \bmod W)$ and $\widetilde{A}(V \bmod W)$, respectively.

Remark. In the PL category, $A(D^n)$ and $\widetilde{A}(D^n)$ are contractible by the Alexander trick. Likewise $A(D^n \bmod D^p)$ and $\widetilde{A}(D^n \bmod D^p)$ are contractible. Hence $E(D^p, D^n; i)$ and $\widetilde{E}(D^p, D^n; i)$ are contractible. This in turn implies $C(D^p, D^n; i)$ and $\widetilde{C}(D^p, D^n; i)$ are contractible.

These results (except for $\widetilde{C}(D^p, D^n; i)$, see 2.1 below) are definitely false in the smooth category [21].

This enables us to define $\pi_i(\widetilde{E}(W, V; g))$ and $\pi_i^{rel}(E(W, V; g)) = \pi_i(\widetilde{E}(W, V; g), E(W, V; g))$, etc. A homotopy class in $\pi_i(\widetilde{E}(W,V; g))$ is represented by a simplex $\varphi: \Delta_i \times W \to \Delta_i \times V$ in $\widetilde{E}(W, V; g)$ such that $\varphi = id \times g$ on $\partial\Delta_i \times W$. Equivalently, we can take an embedding $\varphi: R^i \times W \to R^i \times V$ such that $\varphi = id \times g$ outside a compact set. A homotopy class of $\pi_i^{rel}(E(W, V; g))$ is represented by a $\varphi \in \widetilde{E}(W, V; g)$ such that on $\partial\Delta_i \times W$, φ commutes with projection onto Δ_i, and $\varphi | 0 \times W = id \times g$, $0 \in \Delta_i$ the zero vertex. Two such simplices represent the same homotopy class if there is a smooth (PL) concordance between them preserving these conditions. See Morlet [10] or A1.

Lemma 2.1. $\widetilde{C}(W, V; g)$ is contractible $(\dim V - \dim W \geq 3)$.

Proof. We first indicate the proof schematically. Given φ as above, we define $\Phi: \Delta_i \times W \times I \times I \to \Delta_i \times V \times I \times I$ as indicated by the diagram below:

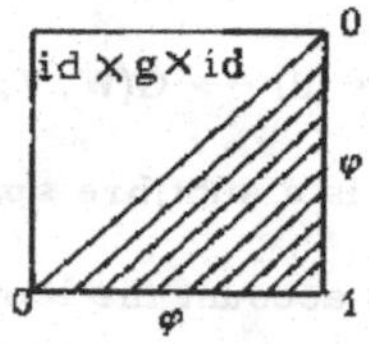

This gives a concordance of φ to $id \times g \times id$ preserving the desired conditions and thus defining a contraction.

We now give the details:

Smooth Category: By an initial isotopy, we may assume $\varphi: \Delta_i \times W \times I \to \Delta_i \times V \times I$ is a product near $\Delta_i \times W \times 0$ and $\Delta_i \times W \times 1$. Define $\lambda: I^2 \to T$ by $\lambda(r, s) = (1 - s)(r, 0) + s(1, 1-r)$. Then $\lambda_s: I \to T$, $\lambda_s(r) = \lambda(r, s)$, is the line from (s, s) to $(1, 0)$. We define Φ on each such line to be φ; i.e. if $\varphi(a, w, r) = (\varphi_1(a, w, r), \varphi_2(a, w, r), \varphi_3(a, w, r))$ we define

$$\Phi(a, w, \lambda(r, s)) = (\varphi_1(a, w, r), \varphi_2(a, w, r), \lambda(\varphi_3(a, w, r), s)) .$$

in $\Delta_i \times W \times T$, and $\Phi = id_{\Delta_i} \times g \times id_{I \times I}$ above the diagonal in I^2.

PL Category: Consider $\varphi \times id: (\Delta_i \times W \times I) \times I \to (\Delta_i \times V \times I) \times I$. Note that $\varphi \times id | \Delta_i \times W \times 0 \times I = id \times g \times id \times id$. Take the isomorphism of $\partial(I^2)$ into itself that sends $0 \times [0, 1/2]$ onto $0 \times [0, 1]$, $0 \times [1/2, 1]$ onto $[0, 1] \times 1$, $[0, 1] \times 1$ onto $1 \times [1, 1/2]$, $1 \times [1, 0]$ onto $1 \times [1/2, 0]$, linearly, and leaves $[0, 1] \times 0$ fixed. Extend this isomorphism to an isomorphism ψ of I^2 onto itself by coning over the center. Then $\Phi = id \times \psi \circ \varphi \times id \circ id \times \psi^{-1}$ defines the desired concordance from φ to $id \times g \times id$.

Thus Φ is a concordance of φ to $\mathrm{id} \times g \times \mathrm{id}$, and on any subcomplex of $\Delta_i \times W \times I$ where φ agrees with $\mathrm{id} \times g \times \mathrm{id}$, Φ is constant. Thus $\tilde{C}(W, V; g)$ is contractible.

Consider the fibre space (dim V - dim W ≥ 3)

$$E(W \times I, V \times I, g \times \mathrm{id}) \longrightarrow C(W, V; g) \longrightarrow E(W, V; g),$$

where $\rho(\varphi) = \varphi | \Delta_i \times W \times 1$. This is a subfibre space of the analogous fibre space for $\tilde{C}(W, V; g)$. Hence taking into account the lemma we get an exact sequence:

$$(2.2) \quad \to \pi_i(C(W, V; g)) \to \pi_{i+1}^{rel}(E(W, V; g)) \to \pi_i^{rel}(E(W \times I, V \times I, g \times \mathrm{id})$$

$$\to \pi_{i-1}(C(W, V; g)) \to \cdots \to \pi_0(C(W, V; g)) \to \pi_1^{rel}(E(W, V; g)) \to 0 .$$

Note: $\pi_0^{rel}(E(W, V; g) = 0$ by definition of $\tilde{E}(W, V; g)$ and $E(W, V; g)$.

Lemma of Disjunction -- 2nd form: Let V be a manifold of dim n, $g: D^p \to V$, $h: D^q \to V$ disjoint embeddings with $n-p \geq 3$, $n-q \geq 3$. Then $\pi_i^{rel}(E(D^p, V - hD^q; g)) \to \pi_i^{rel}(E(D^p, V; g))$ is an isomorphism for $i \leq 2n-p-q-5$ and surjective for $i = 2n-p-q-4$. If V is 1-connected then this holds for $i \leq 2n-p-q-4$ and $i = 2n-p-q-3$, respectively.

Proof. Taking the exact sequences above for $C(D^p, V - hd^q; g)$ and $C(D^p, V; g)$ we get an exact sequence:

$$\to \pi_i(C(D^p, V; g)), C(D^p, V - hD^q; g) \to \pi_{i+1}(\tilde{E}(D^p, V; g), E(D^p, V; g), \tilde{E}(D^p, V - hD^q; g)$$

$$\to \pi_i(\tilde{E}(D^p \times I, V \times I, g \times \mathrm{id})), E(D^p \times I, V \times I, g \times \mathrm{id}), \tilde{E}(D^p \times I, (V - hD^q) \times I, g \times \mathrm{id})$$

$$\to \pi_{i-1}(C(D^p, V; g), C(D^p, V - hD^q; g)) \to .$$

Hence from the first form of the Lemma of Disjunction, we see that for $i \leq 2n-p-q-5$, $\pi_{i+1}(\tilde{E}(D^p, V; g), E(D^p, V; g), \tilde{E}(D^p, V - hD^q; g) = 0$ whenever $\pi_i(\tilde{E}(D^{p+1}, V \times I, g \times \mathrm{id}), E(D^{p+1}, V \times I, g \times \mathrm{id}), \tilde{E}(D^{p+1}, V \times I - D^{p+1}, g \times \mathrm{id})) = 0$. Thus by induction, these groups are zero and the result follows from the exact sequence of the triad. (See remark below.)

Remarks: π_1 of a triad is not usually defined, but in our case it makes sense to say $\pi_1(\widetilde{E}(D^p, V; g), E(D^p, V; g), \widetilde{E}(D^p, V-hD^q;)) = 0$ if $\pi_1^{rel}(E(D^p, V-hD^q; g)) \to \pi_1^{rel}(E(D^p, V; g))$ is surjective. Both of these last groups are zero by (2.2) and the fact that $\pi_0^{rel}(E(D^p, V-hD^q; g)) = \pi_0^{rel}(E(D^p, V; g)) = 0$ and $\pi_0(C(D^p, V-hD^q; g)) = \pi_0(C(D^p, V; g)) = 0$ by Hudson.

In any case, one may prove directly from (2.2) that $\pi_2^{rel}(E(W, V-hD^q; g)) \to \pi_2^{rel}(E(W, V; g))$ is surjective, and hence that π_2 of the triad is zero.

For the rest of this section we assume $n-p \geq 3$.

Lemma 2.3. For all i, there is a natural homomorphism

$$\pi_i^{rel}(E(D^p, V; g)) \to \pi_{i+1}^{rel}(E(D^{p-1}, V; g|D^{p-1}),$$

and this homomorphism is an isomorphism for $i \leq 2n-2p-5$ and surjective for $i = 2n-2p-4$. If V is 1-connected this holds for $i \leq 2n-2p-4$ and $i = 2n-2p-3$ respectively.

Proof. Let $B(D^p, V; g)$ be the space of embeddings of D^p_+ in V, which agree with g on S^{p-1}_+, and define $\widetilde{B}(D^p, V; g)$ similarly. These spaces are contractible since D^p_+ may be shrunk into a collar neighborhood of S^{p-1}_+, where our embeddings can be assumed to agree with g. Further, we have the fibration

$$F(D^p, V; g) \to B(D^p, V; g) \to E(D^{p-1}, V; g|D^{p-1}),$$

where $F(D^p, V; g) \subset B(D^p, V; g)$ are those embeddings of D^p_+ which agree with g on D^{p-1} as well as S^{p-1}_+. Similarly for $\widetilde{B}(D^p, V; g)$. This gives an isomorphism

a) $$\alpha : \pi_i^{rel}(F(D^p, V; g)) \to \pi_{i+1}^{rel}(E(D^{p-1}, V; g|D^{p-1})), \quad \text{all } i.$$

On the other hand, $E(D^p, V; g)$ has the same homotopy type as its subspace $E(D^p, V, g; D^p_-)$; i.e. embeddings which agree with g on D^p_- as well as ∂D^p. Evidently we have a map $\gamma : E(D^p, V, g; D^p_-) \to F(D^p, V; g)$. Similarly for $\widetilde{E}(D^p, V; g)$. Thus we get a homomorphism $\beta : \pi_i^{rel}(E(D^p, V; g)) \to \pi_i^{rel}(F(D^p, V; g)$, such that

$$\begin{array}{ccc} \pi_i^{rel}(E(D^p, V; g)) & \xrightarrow{\ \beta\ } & \pi_i^{rel}(F(D^p, V; g)) \\ \nwarrow{\scriptstyle \simeq} & & \nearrow{\scriptstyle \gamma} \\ & \pi_i^{rel}(E(D^p, V; g; D^p_-)) & \end{array}$$

commutes. In composing α and β we get a homomorphism

b) $$\delta : \pi_i^{rel}(E(D^p, V; g)) \to \pi_{i+1}^{rel}(E(D^{p-1}, V; g|D^{p-1})) \ .$$

Consider the subspace $F_0(D^p, V; g) \subset F(D^p, V; g)$ of embeddings which also coincide with g in a neighborhood of D^{p-1}; and similarly, $\widetilde{F}_0(D^p, V; g)$.

c) There is a homomorphism $\gamma' : \pi_i^{rel}(E(D^p, V, g; D^p_-)) \to \pi_i^{rel}(F_0(D^p, V; g))$ with γ' an isomorphism for $i \leq 2n-2p-5$ and surjective for $i = 2n-2p-4$. Further

$$\begin{array}{ccc} & \xrightarrow{\ \gamma'\ } & \pi_i^{rel}(F_0(D^p, V; g)) \\ \pi_i^{rel}(E(D^p, V, g; D^p_-)) & & \downarrow \\ & \xrightarrow{\ \gamma\ } & \pi_i^{rel}(F(D^p, V; g)) \end{array}$$ commutes.

In fact, delete a tubular neighborhood of $g(D^{p-1})$ in V. What remains is a submanifold V', and the space $F_0(D^p, V; g)$ appears as equivalent to $E(D^p, V'; g')$, where $g'(D^p) = g(D^p_+) \cap V'$. Likewise $E(D^p, V, g; D^p_-)$ appears as equivalent to $E(D^p, V' - gD^p_- \cap V'; g')$. By the 2nd form of the Lemma of Disjunction, $\pi_i^{rel}(E(D^p, V' - gD^p_- \cap V', g') \to \pi_i^{rel}(E(D^p, V', g'))$ is an isomorphism for $i \leq 2n-2p-5$ and surjective for $i = 2n-2p-4$. Thus c) follows.

d) In the case Diff: $\pi_i^{rel}(F_0(D^p, V; g)) \approx \pi_i^{rel}(F(D^p, V; g))$, all i.

In fact, consider the fibration of F (resp $\widetilde{F}$) with base the space of normal vector fields to $g(D^{p-1})$ and with fibre F_0 (resp. $\widetilde{F}_0$). The result d) is immediate.

e) In the case PL: $\pi_i^{rel}(F_0(D^p, V; g)) \to \pi_i^{rel}(F(D^p, V; g))$ is an isomorphism for $i \leq 2n-2p-4$ and surjective for $i \leq 2n-2p-3$.

One has a fibration of $F(D^p, V; g)$ with base space $N(D^{p-1} \times I, D^n, \text{inc.})$ of embeddings of a neighborhood of D^{p-1} in D^n (i.e., a normal tube of D^{p-1} in V), with fibre $F_0(D^p, V; g)$. Similarly for $\widetilde{F}(D^p, V; g)$ we have the fibering with base $\widetilde{N}(D^{p-1} \times I, D^n; \text{inc.})$ and fibre $\widetilde{F}_0(D^p, V; g)$. In particular, we have the fibering: $F_0(D^p, D^n; \text{inc.}) \to F(D^p, D^n; \text{inc.}) \to N(D^{p-1} \times I, D^n; \text{inc.})$. But $\pi_i(F(D^p, D^n; \text{inc.}) \simeq \pi_{i+1}(E(D^{p-1}, D^n; \text{inc.}) = 0$ by the Alexander trick. Similarly, $\pi_i(\widetilde{F}(D^p, D^n; \text{inc.}) = 0$, all i.

On the other hand, by (c), $\pi_i^{rel}(E(D^p, D^n, \text{inc}; D^p_-)) \to \pi_i^{rel}(F_0(D^p, D^n; \text{inc}))$ is surjective for $i \leq 2n-2p-4$, and the first group is again zero by the Alexander trick. Hence $\pi_i^{rel}(F_0(D^p, D^n; \text{inc.})) = 0$ for $i \leq 2n-2p-4$ and $\pi_i^{rel}(N(D^{p-1} \times I, D^n; \text{inc.})) = 0$ for $i \leq 2n-2p-3$. Hence the result.

f) The lemma now follows by a), c), d) and e). If V is 1-connected the same argument applies with i shifted by 1.

Proposition 2.4. If $\dim V = n$ and if $\pi_j(V) = 0$ for $j \leq k$, $\pi_i^{rel}(E(pt, V)) = 0$ for $i \leq n+k-2$ ($k \leq n-2$ in the diff. case).

Proof. An element α of $\pi_i^{rel}(E(pt, V))$ is represented by an embedding $f: D^i \to D^i \times V$ such that

a) $f(y) = (y, v_0)$ for $y \in S_-^{i-1}$, v_0 the base point of V,

b) $f(y) \in y \times V$ for $y \in S_+^{i-1}$.

Now f represents 0 in $\pi_i^{rel}(E(pt, V))$ if there exists an embedding $g: D^i \times I \to D^i \times I \times V$ such that:

1) $g|D^i \times 0 = f$

2) $g(y) = (y, v_0)$ for $y \in S_-^{i-1} \times I \cup D^i \times 1$

3) $g(y) \in y \times V$ if $y \in S_+^{i-1} \times I$.

Obviously, there is a homotopy $h: D^i \times I \to D^i \times I \times V$ satisfying 1), 2), and 3) and with $h|\partial D^i \times I$ an embedding. But then h is homotopic mod $\partial(D^i \times I)$ to an embedding g if $i \leq n+k-2$ in the PL case by Hudson [A1], and if $i \leq n+k-2$, $k \leq n-2$ in the Diff case by Haefliger [A1].

Proposition 2.5. Let g be an embedding of D^p into V^n. If $\pi_j(V) = 0$ for $j \leq k$, then $\pi_i^{rel}(E(D^p, V; g)) = 0$ for $i \leq \inf(2n-2p-4, n+k-p-2)$.

Proof. By Lemma 2.3, if $n-p \geq 3$ and $i \leq 2n-2p-5$ ($i \leq 2n-2p-4$), $\pi_i^{rel}(E(D^p, V; g) \simeq \pi_{i+p}^{rel}(E(pt. V))$, and the result follows from 2.4. (For $k = 0$, $2n-2p-5 \geq n-p-2$ if $n-p \geq 3$.)

Remark. Note $\pi_0^{rel}(E(D^p, V; g)) = 0$ for all p.

Proposition 2.6. With g and V as in 2.5. If $\pi_j(V) = 0$ for $j \leq k$, then $\pi_i(C(D^p, V; g)) = 0$ for $i \leq \inf(2n-2p-5, n-p+k-3)$.

Proof. By Proposition 2.5, and the exact sequence (2.2). In fact, we have

$$\to \pi_{i+1}^{rel}(E(D^{p+1}, V \times I, g \times id)) \to \pi_i(C(D^p, V; g)) \to \pi_{i+1}^{rel}(E(D^p, V; g)) \to .$$

But $i+1 \leq \inf(2n+2-2p-2-4, n+1-p-1+k-2) = \inf(2n-2p-4, n-p+k-2)$ if and only if $i \leq \inf(2n-2p-5, n-p+k-3)$.

By Theorem 1.7, Proposition 2.6, and the Alexander trick (see beginning of this section) we have:

Theorem 2.7. Let V^n, $n \geq 5$, be a k-connected compact PL manifold with non-empty boundary. Then $\pi_i(C(D^p, V; g)) = 0$ for $i \leq n+k-p-3$ and $p \leq n-3$. If $n = 5$ or ∂V not 1-connected we require $k \leq \max(n-4, n-p-2)$.

From 2.7 and the exact sequence (2.2) we have:

Theorem 2.8. With V as in (2.7), $\pi_i^{rel}(E(D^p, V; g)) = 0$ for $i \leq n+k-p-2$ and $p \leq n-3$. If $n = 5$ or ∂V not 1-connected we require $k \leq \max(n-4, n-p-2)$.

In particular, using the remark following Theorem 1.7 we get:

Proposition 2.9. $\pi_i^{rel}(E(D^p, S^{n-p} \times D^p)) = 0$ for $i \leq n+k-p-2 = 2n-2p-3$, $p \leq n-3$, $n \geq 5$.

3. Automorphisms and Concordances

Let V^n be a compact smooth (PL) manifold, $n \geq 5$, and let $A(V)$ and $\widetilde{A}(V)$ be as in Chapter 2, i.e. automorphisms fixed on ∂V.

If $W \subset \text{Int } V$ is a compact submanifold with $\dim W = \dim V$, then since an i-simplex φ in $A(W)$ (resp. $\widetilde{A}(W)$) is the identity on $\Delta_i \times \partial W$, it may be extended to an i-simplex of $A(V)$ (resp. $\widetilde{A}(V)$) which is the identity outside $\Delta_i \times W$. Thus we have an injection $\alpha : (\widetilde{A}(W), A(W)) \to (\widetilde{A}(V), A(V))$. Let $\pi_i^{rel}(A(V)) = \pi_i(\widetilde{A}(V), A(V)) = \pi_i(\widetilde{A}(V)/A(V))$. Let

$$\begin{aligned}\pi_i^{rel}(\alpha) &= \pi_i(\widetilde{A}(V); \widetilde{A}(W), A(V)) \\ &= \pi_i(\widetilde{A}(V)/A(V), \widetilde{A}(W)/A(W)) \\ &= \pi_i(\widetilde{A}(V)/\widetilde{A}(W), A(V)/A(W)) .\end{aligned}$$

Lemma a (Diff Category). Let $V = W \cup$ an $(n-i)$-handle; i.e., $V - \text{Int } W = \partial V \times I \cup$ an i-handle, $i \leq n-3$. Then if V is k-connected, $\pi_j^{rel}(\alpha) = 0$ for $j \leq \inf(2n-2i-4, n+k-i-2)$.

Lemma b (PL Category). Let $V = W \cup$ an $(n-i)$-handle. Then if V is k-connected, $\pi_j^{rel}(\alpha) = 0$ for $j \leq \inf(2n-2i-4, n+k-i-2, n-i+2)$.

Before proving Lemmas a and b we derive the immediate consequences.

Theorem 3.1. Given V^n, $n \geq 5$, and $W^n \subset \text{Int } V^n$. Suppose

1) $\pi_i(\partial W) \simeq \pi_i(W)$, $i = 0, 1$,

2) $\pi_i(V, W) = 0$ for $i \leq r$, $r \leq n-4$ (or $r \leq n-3$ if $n > 5$ and $\pi_i(\partial V) = \pi_i(V)$, $i = 0, 1$),

3) W is k-connected, $0 \leq k \leq r$.

Then $\pi_j^{rel}(\alpha) = 0$ for

a) $j \leq \inf(2r-2, r+k-1)$, Diff Category

b) $j \leq \inf(2r-2, r+k-1, r+3)$, PL Category .

Proof. The hypotheses 1) and 2) imply, $V = W \cup (n-i)$-handles, $n-i > r$; or $V - \text{Int } W = \partial V \times I \cup i$-handles, $0 \leq i \leq n-r-1$. We apply Lemmas a and b inductively to straighten out the handles, starting with $i = 0$. Each time we take $V' = W \cup$ remaining handles, $W' = V'$ minus a single remaining handle of top dimension, and apply the Lemmas to V' and W'. The condition $i \leq n-r-1$ implies a) and b).

Corollary 3.2. Given V^n, $n \geq 5$, and $D^n \subset \text{Int } V$, suppose V k-connected, ($k \leq n-4$ if ∂V not 1-connected or $n = 5$). Then $\pi_j^{rel}(\alpha) = 0$ for

a) $j \leq 2k-2$, Diff Category,

b) $j \leq \inf(2k-2, k+3)$, PL Category.

Remark. If V has a handle decomposition with only handles of $\dim > k$, the condition $k \leq n-4$ in (3.2) may be dropped.

Proof. Take $r = k$ in 3.1. Note that $V = S^n$ or D^n if $k > n-3$ and ∂V is 1-connected. For $V = S^n$ the result follows from Lemmas a and b, taking $i = 0$. For $V = D^n$, $\pi_j^{rel}(\alpha) = 0$ all j, by uniqueness of collars.

Corollary 3.3. Given V^n, $n \geq 5$, and $D^n \subset \text{Int } V$, suppose V is k-connected, $k+1 < n/2$, and the tangent bundle $T(V)$ is trivial over the $(k+1)$-skeleton of V. Then $\alpha_* : \pi_j^{rel}(A(D^n)) \to \pi_j^{rel}(A(V))$ is an isomorphism for

a) $j \leq 2k-1$, Diff Category,

b) $j \leq \text{in}(2k-1, k+3)$, PL Category.

Proof. Let N be a regular neighborhood of the $k+1$-skeleton of V. Now N immerses in D^n and by general position there is an embedding $f: N \to \text{Int } D^n$, and $\pi_j(f) = 0$ for $j \leq k+1$. If $\bar{\alpha}: \tilde{A}(N) \to \tilde{A}(V)$ and $\beta: \tilde{A}(N) \to \tilde{A}(D^n)$ are the injections, $\pi_j^{rel}(\bar{\alpha}) = \pi_j^{rel}(\beta) = 0$ for $j \leq 2k$ ($j \leq \inf(2k, k+4)$, PL case) by (3.1). Since $D^n \subset N \xrightarrow{f} D^n$ induces an isomorphism $\pi_j^{rel}(A(D^n)) \to \pi_j^{rel}(A(D^n))$ for all j, $\pi_j^{rel}(A(D^n)) \to \pi_j^{rel}(A(N))$ is an isomorphism for $j \leq 2k-1$ ($j \leq \inf(2k-1, k+3)$

PL case). Hence $\alpha_* : \pi_j^{rel}(A(D^n)) \to \pi_j^{rel}(A(V))$ is an isomorphism for $j \leq 2k-1$ ($j \leq \inf(2k-1, k+3)$ PL case).

Remark. a) Diff Category. If $k \equiv 2, 4, 5, 6 \pmod 8$, then if V is k-connected, TV is always trivial over the (k+1)-skeleton.

b) PL Category. If $k = 2, 4, 5, 6$, then if V is k-connected, TV is always trivial over the (k+1)-skeleton.

c) PL Category. $\pi_j^{rel}(A(D^n)) = 0$ all j by Alexander's trick.

Proofs of Lemmas a and b. Let $h: (D^i \times D^{n-i}, S^{i-1} \times D^{n-i}) \to (V, \partial V)$ be the i-handle. Identifying R^{n-i} with $\mathrm{Int}\, D^{n-i}$, we have the fibrations

(3.3) $$A(V; D^i) \to A(V) \to E(D^i, V; h)$$

(3.4) $$A(W) \to A(V; D^i) \to A_\gamma(D^i \times R^{n-i}; D^i)$$

and similarly for $\widetilde{A}$; where

$A(V; D^i)$ = automorphisms of V fixed on $h(D^i \times 0)$

$A_\gamma(D^i \times R^{n-i}; D^i)$ = germs of embeddings of $D^i \times R^{n-i}$ in $D^i \times R^{n-i}$ fixed on $\partial D^i \times R^{n-i}$ and in $D^i \times 0$.

In (3.4) we have substituted $A_\gamma(D^i \times R^{n-i}; D^i)$ for $E(D^i \times R^{n-i}, V, h; D^i \times 0)$ since they are obviously of the same homotopy type.

From (3.3) for A and $\widetilde{A}$, we get the exact sequence:

(3.3)' $$\to \pi_j^{rel}(A(V, D^i)) \to \pi_j^{rel}(A(V)) \to \pi_j^{rel}(E(D^i, V; h) \to \pi_{j-1}^{rel}(A(V, D^i) \to$$

and from (3.4), the exact sequence

(3.4)' $$\to \pi_j^{rel}(A(W)) \to \pi_j^{rel}(A(V; D^i)) \to \pi_j^{rel}(A_\gamma(D^i \times R^{n-i}; D^i)) \to \pi_{j-1}^{rel}(A(W)) \to .$$

We first complete the proof in the Diff case.

Diff Case. $A_\gamma(D^i \times R^{n-i}; D^i)$ and $\widetilde{A}_\gamma(D^i \times R^{n-i}; D^i)$ both have the homotopy type of the complex of vector bundle equivalences of $D^i \times R^{n-i}$ fixed on $\partial D^i \times R^{n-i}$. Hence $\pi_j^{rel}(A_\gamma(D^i \times R^{n-i}; D^i) = 0$, all j. Hence $\pi_j^{rel}(A(W)) \simeq \pi_j^{rel}(A(V; D^i))$, all j.

On the other hand, $\pi_j^{rel}(E(D^i, V; h) = 0$ for $j \leq \inf(2n-2i-4, n+k-i-2)$ by (2.5). Hence the result follows from (3.3)'.

PL Case. Taking $V = D^n$ and hence $W = S^{n-i-1} \times D^{i+1}$ in (3.4), we have from (3.4)', using the fact that $A(D^n; D^i)$ and $\widetilde{A}(D^n; D^i)$ are contractible by the Alexander trick:

(3.5) $$\pi_j^{rel}(A_\gamma(D^i \times R^{n-i}; D^i)) \simeq \pi_{j-1}^{rel}(A(S^{n-i-1} \times D^{i+1}) .$$

Taking $V = S^{n-i} \times D^i$ and hence $W = D^n$ in (3.3) and (3.4), and using the fact that $\pi_j^{rel}(E(D^i, V; h) = 0$ for $j \leq n+k-i-2$ by (2.9), we get from (3.3)' and (3.4)' since $A(D^n)$ and $\widetilde{A}(D^n)$ are trivial:

(3.6) $\pi_j^{rel}(A_\gamma(D^i \times R^{n-i}; D^i)) \to \pi_j^{rel}(A(S^{n-i} \times D^i))$ is an isomorphism for

$j < n+k-i-2 = 2n-2i-3$.

From (3.5) and (3.6) we have

(3.7) $$\pi_j^{rel}(A_\gamma(D^i \times R^{n-i}; D^i)) \simeq \pi_{i+j}^{rel}(A_\gamma(R^n, 0)), \quad j \leq 2n-2i-4$$

Now $\pi_j^{rel}(A_\gamma(R^n, 0)) = \pi_j(\widetilde{PL}_n, PL_n) = 0$ for $j \leq n+2$ by [3]. By (3.7), $\pi_j^{rel}(A_\gamma(D^i \times R^{n-i}; D^i) \simeq 0$ for $j \leq \inf(2n-2i-4, n-i+2)$. Applying (3.3)' and (3.4)', we get Lemma b using (2.5).

From Corollaries 3.2 and 3.3 we get:

Theorem A. Let V^n be a compact k-connected manifold, $n \geq 5$, and $D^n \subset \text{Int } V^n$.

a) Diff Category: $\pi_j(A(V), A(D^n)) \to \pi_j(A(V), A(D^n))$ is an isomorphism if

1. $j \leq 2k-3$ ($k \leq n-4$ if ∂V not 1-connected or $n = 5$)

2. $j \leq 2k-2$ if $k+1 < n/2$ and either $k \equiv 2,4,5,6 \pmod 8$ or TV is trivial over the $(k+1)$-skeleton of V.

b) PL Category: $\pi_j(A(V)) \to \pi_j(\tilde{A}(V))$ is an isomorphism if

1. $j \leq \inf(2k-3, k+2)$ ($k \leq n-4$ if ∂V not 1-connected or $n = 5$)
2. $j \leq 2k-2$ if $k+1 < n/2$ and either $k = 2,4$ or $k = 3$ and TV is trivial over the 4-skeleton of V.

Remark. a) Diff Category: $\pi_j(\tilde{A}(V), \tilde{A}(D^n))$ is determined up to a group extension by the homotopy type of V (see [1] or [9]). Hence the same is true of $\pi_j(A(V), A(D^n))$ for the values of j above.

b) PL Category: $\pi_j(\tilde{A}(V))$ is determined up to a group extension by the homotopy type of V (loc. cit.). Hence the same is true of $\pi_j(A(V))$ for the values of j above.

We now investigate the dependence of $\pi_j^{rel}(A(V))$ on the tangential homotopy type of V.

Definition. If X and Y are connected CW complexes, they will be called r-equivalent if the rth stages X_r and Y_r of their Postnikov systems are homotopy equivalent.

Thus there are maps $\rho : X \to X_r$, $\sigma : Y \to Y_r$ and $\varphi : X_r \to Y_r$; where φ is a homotopy equivalence, and ρ, σ induce isomorphisms on π_i, $i \leq r$, and $\pi_i(X_r) \cong \pi_i(Y_r) = 0$ for $i > r$.

Note that if $X^{(r+1)}$ is the $(r+1)$-skeleton of X, then $\rho | X^{(r+1)} : X^{(r+1)} \to X_r$ induces an isomorphism on π_i, $i \leq r$. On the other hand, if $Y^{(r+1)}$ is the $(r+1)$-skeleton of Y_r, there exists a map $\lambda : Y_r^{(r+1)} \to Y$ such that $\sigma\lambda : Y_r^{(r+1)} \to Y_r$ is homotopic to the inclusion. Now $\rho | X^{(r+1)}$ is homotopic to $\rho' : X^{(r+1)} \to X_r^{(r+1)}$, and we may assume φ sends $X_r^{(r+1)}$ into $Y_r^{(r+1)}$. Then $\lambda\varphi\rho' : X^{(r+1)} \to Y$ induces an isomorphism on π_i, $i \leq r$. By adding a wedge of $(r+1)$-spheres to

$X^{(r+1)}$, we get an (r+1)-complex $K \supset X^{(r+1)}$ such that $\lambda\varphi\rho'$ extends to a map $g: K \to Y$ with $\pi_i(g) = 0$ for $i \leq r+1$. Also the inclusion $i: X^{(r+1)} \to X$ may be extended to $f: K \to X$ by sending the extra (r+1)-spheres to the base point, so that $\pi_i(f) = 0$ for $i \leq r+1$. Hence we have shown:

Lemma 3.8. If X and Y are r-equivalent, there exists an (r+1)-complex K and maps $f: K \to X$, $g: K \to Y$ with $\pi_i(f) = \pi_i(g) = 0$ for $i \leq r+1$. Further, $g_* f_*^{-1} = \sigma_*^{-1} \varphi_* \rho_*$ on $\pi_i(x)$, $i \leq r$.

Corollary 3.9. An r-equivalence determines a unique identification of $H^i(X)$ and $H^i(Y)$, $i \leq r$, such that $f^* = g^*$.

Remarks. 1. If $X^{(r+1)}$ and $Y^{(r+1)}$ are homotopy equivalent, then X and Y are r-equivalent.

2. If X is r-connected, then X is r-equivalent to any contractible Y.

Definition. Two n-manifolds V and V' are said to be of the same tangential r-type, if they are r-equivalent and if $f^*TV \simeq g^*TV'$, $f: K \to V$, $g: K \to V'$ as in 3.8.

Remark. V and V' are of the same tangential 2-type, if they are of the same 2-type and $f^* w_i = g^* w_i$, $i = 1, 2$, w_i the Stiefel-Whitney class.

Theorem B. Suppose V, V' are k-connected compact manifolds of dim $n \geq 5$, and that V and V' have the same tangential r-type, $n/2 > r+1 \geq k$. Then $\pi_j^{rel}(A(V) \simeq \pi_j^{rel}(A(V'))$

a) Diff Category: If $j \leq \inf(2r-1, r+k-1)$

b) PL Category: If $j \leq \inf(2r-1, r+k-1, r+3)$.

Proof. By definition, there exists an r+1 complex K and maps $f: K \to V$, $f': K \to V'$ with $\pi_i(f) = \pi_i(f') = 0$ for $i \leq r+1$ and such that $f^*TV = (f')^* TV'$. Since $r+1 < n/2$, we may assume f and f' are embeddings.

Let N and N' be regular neighborhoods of f(K) and f(K'). Then we have a homotopy equivalence $\varphi : N \to N'$ which is covered by a bundle map $T(\varphi): TN \to TN'$, and hence we may assume φ is an immersion of N in Int N'. By general position we may assume φ is an embedding of N in Int N'. Let $\beta : \widetilde{A}(N) \to \widetilde{A}(N')$ be the injection induced by φ, and let $\alpha : \widetilde{A}(N) \to \widetilde{A}(V)$, $\alpha' : \widetilde{A}(N') \to \widetilde{A}(V')$ be the injections. By (3.1), $\pi_j^{rel}(\alpha) = \pi_j^{rel}(\alpha') = \pi_j^{rel}(\beta) = 0$ for $j \leq \inf(2r, r+k)$ Diff case or $j \leq \inf(2r, r+k, r+4)$ PL case. Hence $\pi_j^{rel}(A(V)) \simeq \pi_j^{rel}(A(N)) \simeq \pi_j^{rel}(A(N')) \simeq \pi_j^{rel}(A(V'))$ for j as required.

We now apply these results on A(V) to get results on the group C(V) of concordances of V, C(V) as in section 2; i.e. concordances trivial on ∂V.

If $W \subset \text{Int } V$ is a compact submanifold with dim W = dim V, there is an injection $\gamma: C(W) \to C(V)$ by taking $\gamma\varphi$ to be the identity outside $W \times I$.

Now the exact sequence (2.2) when applied to C(V) gives the exact sequence

$$(3.10) \qquad \to \pi_{i+1}^{rel}(A(V)) \to \pi_i^{rel}(A(V \times I)) \to \pi_{i-1}(C(V)) \to \pi_i^{rel}(A(V)) \to \cdots$$
$$\to \pi_1^{rel}(A(V \times I)) \to \pi_0(C(V)) \to \pi_1^{rel}(A(V)) \to 0 .$$

The sequence (3.10) for W maps into the sequence for V. In fact, letting $\alpha : \widetilde{A}(W) \to \widetilde{A}(V)$, and $\overline{\alpha} : \widetilde{A}(W \times I) \to \widetilde{A}(V \times I)$, ($W \times I$ can be considered in Int $V \times I$, since we may assume our automorphisms in $\widetilde{A}(W \times I)$ are fixed on an ϵ-neighborhood of $W \times \partial I$ and hence we may remove $W \times ([0, \epsilon) \cup (1-\epsilon, 1])$ from $W \times I$), we get a relative exact sequence:

$$(3.11) \qquad \to \pi_{i+1}^{rel}(\alpha) \to \pi_i^{rel}(\overline{\alpha}) \to \pi_{i-1}(\gamma) \to \pi_i^{rel}(\alpha) \to \cdots .$$

Applying 3.1 to α and $\overline{\alpha}$, noticing that $V \times I$ and $W \times I$ satisfy the conditions of 3.1 whenever V and W do, we get from (3.11) the result:

Theorem 3.1'. Given V^n, $n \geq 5$, and $W^n \subset \operatorname{Int} V^n$. Suppose

1. $\pi_i(\partial W) \simeq \pi_i(W)$, $i = 0, 1$,
2. $\pi_i(V, W) = 0$ for $i \leq r$, $r \leq n-4$ (or $r \leq n-3$ if $n > 5$ and $\pi_i(\partial V) \simeq \pi_i(V)$, $i = 0, 1$),
3. W is k-connected, $0 \leq k \leq r$.

Then $\pi_j(\gamma) = 0$ for

a) $j \leq \inf(2r-3, r+k-2)$, Diff Category

b) $j \leq \inf(2r-3, r+k-2, r+2)$, PL Category.

Corollary 3.2'. Given V^n, $n \geq 5$, and $D^n \subset \operatorname{Int} V$. Suppose V is k-connected ($k \leq n-4$ if ∂V is not 1-connected). Then $\pi_j(\gamma) = 0$ for

a) $j \leq 2k-3$, Diff Category,

b) $j \leq \inf(2k-3, k+2)$, PL Category.

Corollary 3.3'. Given V^n, $n \geq 5$, and $D^n \subset \operatorname{Int} V$. Suppose V is k-connected, $k+1 < n/2$, and the tangent bundle $T(V)$ is trivial over the $(k+1)$-skeleton of V. Then $\gamma: \pi_j(C(D^n)) \to \pi_j(C(V))$ is an isomorphism for

a) $j \leq 2k-2$, Diff Category,

b) $j \leq \inf(2k-2, k+2)$, PL Category.

Proof. Theorem 3.1' and Corollary 3.2' follow immediately from 3.1, Corollary 3.2 and (2.19). Corollary 3.3' follows from 3.1' by exactly the same argument by which we derived 3.3 from 3.1, except we use $\pi_j C(V)$ in place of $\pi_j^{rel}(A(V))$, $C(N)$ in place of $\widetilde{A}(N)$, etc.

From Corollaries 3.2', 3, 3' we get:

<u>Theorem A'.</u> Let V^n be a compact k-connected manifold, $n \geq 5$, and $D^n \subset \text{Int } V^n$.

a) <u>Diff Category.</u> $\gamma_* : \pi_j(C(D^n)) \to \pi_j(C(V))$ is an isomorphism for

1. $j \leq 2k-4$ ($k \leq n-4$ if ∂V not 1-connected or $n = 5$),
2. $j \leq 2k-2$ if $k \leq n-4$ ($k+1 < n/2$) and either $k \equiv 2, 4, 5, 6 \pmod 8$ or TV is trivial over the (k+1)-skeleton of V.

b) <u>PL Category.</u> $\pi_j(C(V)) = 0$ if

1. $j \leq \inf(2k-3, k+2)$ ($k \leq n-4$ if ∂V not 1-connected or $n = 5$),
2. $j \leq 2k-2$ if $k+1 < n/2$ and either $k = 2, 4$, or $k = 3$ and TV is trivial over the 4-skeleton of V.

<u>Proof.</u> We need only note that $C(D^n)$ is contractible in the PL Category, by the Alexander trick.

In the same way that Theorem B follows from 3.1, Theorem B' below follows from 3.1'.

<u>Theorem B'.</u> Suppose V, V' are k-connected compact manifolds of dim $n \geq 5$, and that V and V' have the same tangential r-type, $n/2 > r+1 \geq k$. Then $\pi_j(C(V)) \simeq \pi_j(C(V'))$

a) <u>Diff Category.</u> If $j \leq \inf(2r-2, r+k-2)$;

b) <u>PL Category.</u> If $j \leq \inf(2r-2, r+k-2, r+2)$.

<u>Remark.</u> Hatcher and Wagoner [14] have shown for $n \geq 5$, that $\pi_0(C(V))$ in the Diff Category depends only on $\pi_1(V)$, $\pi_2(V)$ and the action of $\pi_1(V)$ on $\pi_2(V)$. In particular, Cerf has shown [4] that $\pi_0(C(V)) = 0$ for V 1-connected, $n \geq 5$.

Hatcher and Wagoner, and also Volodin [13], have announced that $\pi_1(C(V)) \simeq Z_2 + Wh_3(0) \neq 0$ if V is 2-connected, in the Diff Category, $n \geq 7$. This contrasts with Theorem A'(b), which shows that in the PL Category, $\pi_1(C(V)) = 0$ for V 2-connected.

On the other hand, we have

Theorem C' PL Category: $\pi_0(C(V))$ is the same function of $\pi_1(V)$, $\pi_2(V)$ and the action as in the Diff Category, $n \geq 7$.

Proof. Let N be a regular neighborhood of the 3-skeleton of V. Then by Theorem 3.1', $\pi_j(C(N)) \to \pi_j(C(V))$ is an isomorphism for $j = 0$. Now N is a smoothable PL manifold [8], and hence by [2], $\pi_0(C^d(N)) \simeq \pi_0(C^{p\ell}(N))$. Thus $\pi_0(C^d(N)) \simeq \pi_0(C^{p\ell}(V))$. Since $\pi_i(N) \to \pi_i(V)$ is an isomorphism for $i \leq 2$, and the isomorphisms commute with the action of π_1, the result follows.

Remark. Rourke [11] has outlined an argument that $\pi_0(C(V)) = 0$ in the PL Category when V is 1-connected, $n \geq 5$.

Remark. By definition, $\pi_0(C(V)) \simeq \pi_1^{rel}(A(V))$. Hence the above remarks and Theorem C' apply equally to $\pi_1^{rel}(A(V))$.

The effect of multiplication with S^1.

Let X and Y be compact topological k-ads and consider the Δ-set $Iso(X \times R, Y \times R)$; i.e. an i-simplex is a homeomorphism $f: \Delta_i \times X \times R \to \Delta_i \times Y \times R$ commuting with projection on Δ_i. Further we will assume f is "end-preserving", i.e. f sends the positive end of $\Delta_i \times X \times R$ into the positive end of $\Delta_i \times Y \times R$. If $f_0: X \times R \to Y \times R$ is a base point, we can identify $Iso(X \times R, Y \times R)$ with $A(Y \times R)$ by the correspondence $g \to g \circ id_{\Delta_i} \times f_0$, $g \in A(Y \times R)^{(i)}$.

If X and Y are PL or smooth manifold k-ads and $f_0: X \times R \to Y \times R$ is an isomorphism we define $Iso(X \times R, Y \times R, f_0)$ by identifying it with $A(Y \times R)$ as above. We define $Iso(X \times S^1, Y \times S^1)$ in the Top, PL or smooth categories simi similarly.

If K is an ordered finite simplicial complex and $\varphi: K \to Iso(X \times R, Y \times R)$ is a Δ-map, then we let $\varphi_\#: K \times X \times R \to K \times Y \times R$ be the evaluation map. $\varphi_\#$ is a homeomorphism which commutes with projection on K. (In the smooth case, $\varphi_\#$ is smooth over each simplex of K.

Lemma 3.12. Let $\varphi: K \to \mathrm{Iso}(X \times R, Y \times R)$ be a Δ-map. Then there is a Δ-map $\bar{\varphi}: K \to \mathrm{Iso}(X \times S^1, Y \times S^1)$ such that:

1. On some interval $[0, d]$,

$$\begin{array}{ccc} K \times X \times [0,d] & \xrightarrow{\varphi_\#} & K \times Y \times R \\ \downarrow \mathrm{id} \times \exp & & \downarrow \mathrm{id} \times \exp \\ K \times X \times S^1 & \xrightarrow{\bar{\varphi}_\#} & K \times Y \times S^1 \end{array}, \text{ commutes.}$$

2. Let $A = \{(k,x) \in K \times X \mid \pi \circ \varphi_\#(k,x,t) = t\text{ , all } t\}$, $\pi: K \times X \times R \to R$ the projection. (I.e. $\varphi_\# | A \times R = g \times 1_R$, $g: A \to K \times Y$.) Then $\bar{\varphi}_\# | A \times S^1 = g \times \lambda$, where $\lambda: S^1 \to S^1$ is an isomorphism (not depending on $a \in A$).

Remark. a) Here exp means the quotient map $R \to R/Z$ with respect to fixed (Top, Pl or Smooth) actions of Z on R.

b) Note by (1), λ is isotopic to id_{S^1}.

Before proving Lemma 3.12 we note the following:

Corollary 3.13. Let $\varphi: K \to C(V \times R)$ be a Δ-map. Then there is a Δ-map $\bar{\varphi}: K \to C(V \times S^1)$ such that if for some subcomplex $K_0 \subset K$, $\varphi | K_0 = j \circ \varphi_0$, $\varphi_0: K \to C(V)$, $j: C(V) \to C(V \times R)$ the natural injection; then $\bar{\varphi} | K_0 = \bar{j} \circ \varphi_0$, $\bar{j}: C(V) \to C(V \times S^1)$ the natural injection.

Proof of Corollary 3.13. $C(V \times R) = A(V \times I \times R \bmod V \times 0 \times R \cup \partial V \times I \times R)$. The Lemma implies there exists $\bar{\varphi}: K \to A(V \times I \times S^1)$ such that $\bar{\varphi}_\# | K \times V \times 0 \times S^1 \cup K \times \partial V \times I \times S^1 = \mathrm{id} \times \lambda$ and $\bar{\varphi}_\# | K_0 \times V \times I \times S^1 = (\varphi_0)_\# \times \lambda$. Since λ is isotopic to id_{S^1}, we may deform $\bar{\varphi}$ so that $\bar{\varphi}(K) \subset C(V \times S^1) \subset A(V \times I \times S^1)$ and $\bar{\varphi} | K_0 = \bar{j} \circ \varphi_0$. This proves the corollary.

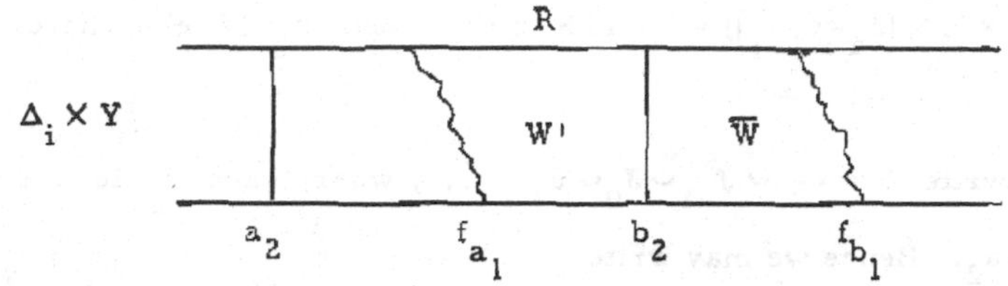

FIGURE 1

Proof of Lemma 3.12. Choose intervals $[a_1, b_1]$ and $[a_2, b_2]$ such that $\varphi_{\#}(K \times X \times a_1) \subset K \times Y \times (a_2, b_2)$, and $\varphi_{\#}(K \times X \times (a_1, b_1)) \supset K \times Y \times b_2$. This is possible since K, X and Y are compact.

Now let $f: \Delta_i \times X \times R \to \Delta_i \times Y \times R$ be a simplex in $\varphi(K)$. We will work with the single simplex f, but since our construction is canonical it will be obvious that the result will agree on faces and define a Δ-map $\overline{\varphi}$.

$$\begin{aligned} \text{Let } W &= \Delta_i \times Y \times [a_2, \infty) - f(\Delta_i \times X \times (a_1, \infty)) \\ W' &= f(\Delta_i \times X \times [a_1, \infty)) - \Delta_i \times Y \times (b_2, \infty) \\ \overline{W} &= \Delta_i \times Y \times [b_2, \infty) - f(\Delta_i \times X \times (b_1, \infty)) . \end{aligned}$$

(See Figure 1) We claim there is an isomorphism $\rho: \overline{W} \to W$. In fact, let $V = \Delta_i \times Y \times [a_2, b_2] \cup \overline{W} = W \cup f(\Delta_i \times X \times [a_1, b_1])$. Now $\partial_{b_2} \overline{W}$ has a collar neighborhood in Int $\overline{W}$; i.e., $\Delta_i \times Y \times [b_2, b_2+\epsilon]$, where ϵ may be chosen independent of the simplex by compactness. Choose an isomorphism $\mu_0: [b_2, b_2+\epsilon] \to [a_2, b_2+\epsilon]$ such that, say $\mu_0 | [b_2+\epsilon/2, b_2+\epsilon] = \text{id}$, and μ_0 is translation on a neighborhood of b_2. Define an isomorphism $\mu: \overline{W} \to V$ by $\mu | \Delta_i \times Y \times [b_2, b_2+\epsilon] = \text{id} \times \mu_0$ and $\mu = \text{id}$ elsewhere. Similarly $\partial_{a_1} W$ has a collar in Int W; i.e., $f(\Delta_i \times X \times [a_1-\epsilon, a_1])$, again with ϵ independent of the simplex. Choose $\nu_0: [a_1-\epsilon, a_1] \to [a_1-\epsilon, b_1]$, an isomorphism with $\nu_0 | [a_1-\epsilon, a_1-\epsilon/2] = \text{id}$ and ν_0 translation near a_1. Define an isomorphism

$\nu: W \to V$ by $\nu | f(\Delta_i \times X \times [a_1 - \epsilon, a_1]) = f \circ \mathrm{id} \times \nu_0 \circ f^{-1}$ and $\nu = \mathrm{id}$ elsewhere. Let $\rho = \nu^{-1}\mu$.

Now we may write $R = \cdots \cup J_{-1} \cup J_0 \cup J_1 \cup \ldots$, where each J_j is an interval of length $b_2 - a_2$. Hence we may write $\Delta_i \times Y \times R = \ldots \cup W_{-1} \cup W'_{-1} \cup W_0 \cup W'_0 \cup W_1 \cup W'_1 \cup \ldots$, where each W_j (W'_j) is a copy of W (W').

Similarly we may write $R = \ldots \cup I_{-1} \cup I_0 \cup I_1 \cup \ldots$, where each I_j is an interval of length $b_1 - a_1$. Define $f': \Delta_i \times X \times R \to \Delta_i \times Y \times R$ by $f' | \Delta_i \times X \times I_j : \Delta_i \times X \times I_j \to \Delta_i \times X \times [a_1, b_1] \xrightarrow{f} W' \cup \overline{W} \xrightarrow{1 \cup \rho} W' \cup W \to W'_j \cup W_{j+1}$. Let $\tau_1, \tau_2 : R \to R$, $\tau_1 : I_j \to I_{j+1}$, $\tau_2 : J_j \to J_{j+1}$. Then f' commutes with the actions and defines $\overline{f}: X \times S^1 \to Y \times S^1$.

If we identify say I_0 with $[a_1, b_1]$ and J_0 with $[a_2, b_2]$ and take $a_1 = 0$ and d sufficiently small so that $f(\Delta_i \times X \times [0, d]) \subset W'$ for all simplices in K, then 1) follows.

If $B = A \cap \Delta_i \times X$ and hence $f | B \times R = g | B \times 1_R$, then $\nu | f(B \times [a_1 - \epsilon, a_1]) = \mathrm{id} \times \nu_0$ and $\rho | g(B) \times [b_2, b_1] = \mathrm{id} \times \lambda'$, $\lambda': [b_2, b_1] \to [a_2, a_1]$. Hence $f' | B \times I_j = g \times \lambda''_j$, $\lambda''_j : I_j \to J_j$ being λ''_0 translated to I_j. Hence $\overline{f} | B \times S^1 = (g|B) \times \lambda$. This completes the proof of Lemma 3.12.

<u>Lemma 3.14.</u> Let K be an ordered finite simplicial complex and $\varphi: K \to C(V)$ a Δ-map. Then if $j: C(V) \to C(V \times R)$ is the natural injection, $j \circ \varphi: K \to C(V \times R)$ is homotopically trivial. (I.e. $j \circ d$ is homotopic to $\varphi': K \to C(V \times R)$, $\varphi'_{\#} = \text{identity}$.)

<u>Proof.</u> Let $f: V \times I \to V \times I$ be a concordance. We will define an isotopy of $f \times 1: V \times I \times R \to V \times I \times R$ to the identity map. Because our construction will be canonical, it will give the desired deformation of $j \circ \varphi$.

We may assume by uniqueness of collares that f is a product on $V \times [0,\epsilon]$ and $V \times [1-\epsilon, 1]$. Write $f(v,t) = (f_1(v,t), f_2(v,t))$. Consider the isotopy $F_s: V \times I \times R \to V \times R \times R$:

$$F_s(v,t,r) = (f_1(v,t+r), f_2(v,t+r) + sg(v,t+r) - r, -sg(v,(t+r)) + r),$$

where $g(v,t+r) = t+r - f_2(v,t+r)$ and $0 \leq s \leq 1$. Here we consider that $f(v,t) = (v,t)$ for $t \leq 0$ and $f(v,t) = (f_1(v,1),t)$ for $t \geq 1$. Then F_s is one-to-one since the sum of the last two coordinates is $f_2(v,t+r)$, and since f is an homeomorphism. This determines v and $t+r$, hence r and therefore t. F_s is an embedding by invariance of domain.

We claim that for $|r| < \epsilon$, $F_s(v,t+r) \subset V \times I \times R$. First for $|r| < \epsilon$ and $t \in [0,1]$, $f_2(v,t+r) - r \in [0,1]$. In fact, if $\epsilon < t+r < 1-\epsilon$, $\epsilon < f_2(v,t+r) < 1-\epsilon$. But if $t+r \leq \epsilon$, $f_2(v,t+r) = t+r$; and if $t+r \geq 1-\epsilon$, $f_2(v,t+r) = t+r$. Thus $f_2(v,t+r) + sg(v,t+r) - r = (1-s)(f_2(v,t+r) - r) + st \in [0,1]$. Since for $|r| < \epsilon$, $F_s(v,0,r) = (v,0,r)$ and $F_s(v,1,r) = (f_1(v),1),1,r)$, F_s is a concordance.

Now $F_0(v,t,r) = (f_1(v,t+r), f_2(v,t+r) - r, r)$ and $F_0(v,t,0) = (f_1(v,t), f_2(v,t), 0)$. By uniqueness of collars, $f \times 1$ is isotopic to F_0. But $F_1(v,t,r) = (f_1(v,t+r), t, f_2(v,t+r) - t)$ is an isotopy and hence is itself isotopic to $(f_1(v,r), t, f_2(v,r)) = (v,t,r)$ for $|r| < \epsilon$. Now by identifying R with $(-\epsilon, \epsilon)$, this gives the desired deformation, at least in the Top or Smooth Category.

In the PL category, we have the problem that multiplying by s does not give a PL-map. To overcome this let $h: I \times [-1,1] \to R$ be a PL approximation to $(s,x) \to sx$ satisfying $h(0,x) = 0$, $h(1,x) = x$, $h(s,0) = 0$. Then substituting $h(s, g(v,t+r))$ for $sg(v,t+r)$ in F_s gives a PL isotopy from F_0 to F_1 as above, and the result follows in this case also.

<u>Theorem D.</u> Let $\varphi: K \to C(V)$ be a Δ-map, K a finite ordered simplicial complex. Then if $\bar{j}: C(V) \to C(V \times S')$ is the natural injection, $\bar{j} \circ \varphi$ is homotopically trivial. (I.e., $\bar{j} \circ \varphi$ is homotopic to $\bar{\varphi}': K \to C(V \times S')$, $\bar{\varphi}'_{\#}$ = identity.)

<u>Proof.</u> By Lemma 3.14, there exists $F: K \times I \to C(V \times R)$ such that $F_0 = j \circ \varphi$ and $F_1 = \varphi'$, $\varphi'_{\#}$ = identity. By the corollary of Lemma 3.12, applied to F, there is a Δ-map $\bar{F}: K \times I \to C(V \times S')$ such that $\bar{F}_0 = \bar{j} \circ \varphi$ and $\bar{F}_1 = \bar{\varphi}'$. That is, $\bar{\varphi}' = j \circ \varphi''$, $\varphi'': K \to C(V)$, $\varphi''_{\#}$ = identity, and $\bar{F}_1 = \bar{j} \circ \varphi'' = \bar{\varphi}'$.

<u>Corollary 1.</u> $\pi_r(\tilde{A}(V)/A(V)) \to \pi_r(\tilde{A}(V \times T^r)/A(V \times T^r))$ is trivial.

<u>Proof.</u> Consider the map of fibrations:

$$\begin{array}{lccccc}
 & \tilde{A}(V \times I)/A(V \times I) & \to & \tilde{C}(V)/C(V) & \to & \tilde{A}(V)/A(V) \\
 & \downarrow & & \downarrow & & \downarrow \\
\text{I} & \tilde{A}(V \times I \times S')/A(V \times I \times S') & \to & \tilde{C}(V \times S')/C(V \times S') & \to & \tilde{A}(V \times S')/A(V \times S') \\
 & \downarrow & & \downarrow & & \downarrow \\
 & \tilde{A}(V \times I \times T^2)/A(V \times I \times T^2) & \to & \tilde{C}(V \times T^2)/C(V \times T^2) & \to & \tilde{A}(V \times T^2)/A(V \times T^2) \\
 & \downarrow & & \downarrow & & \downarrow \\
 & \vdots & & \vdots & & \vdots
\end{array}$$

Since $\tilde{C}(V)$ is contractible, $\tilde{C}(V)/C(V) \cong BC(V)$. By the above theorem $\pi_i BC(V \times T^j) \to \pi_i BC(V \times T^{j+1})$ is trivial, $j \geq 0$, all i. The corollary follows from I and induction.

<u>Corollary 2.</u> Let $j_s: \tilde{A}(V) \to \tilde{A}(V \times T^s)$ be the natural injection and $\bar{j}_s: B\tilde{A}(V) \to B\tilde{A}(V \times T^s)$ the induced map on universal base spaces. Let $\varphi: K \to B\tilde{A}(V)$ be a Δ-map, K a finite ordered simplicial complex with $\dim K \leq r+1$. Then if $n = r(r+1)/2$, there exists a Δ-map $\psi: K \to BA(V \times T^n)$ such that

$$\begin{array}{ccc}
K & \xrightarrow{\psi} & BA(V \times T^n) \\
{\scriptstyle\varphi}\downarrow & & \downarrow{\scriptstyle \bar{i}_n} \\
B\tilde{A}(V) & \xrightarrow{\bar{j}_n} & B\tilde{A}(V \times T^n)
\end{array}$$

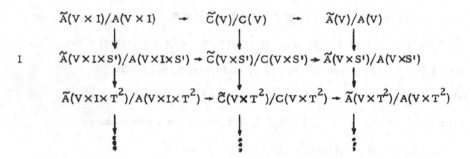

commutes up to homotopy, $\bar{i}_n$ induced by the inclusion $i_n : A(V \times T^n) \to \widetilde{A}(V \times T^n)$.

Proof. Let $\widetilde{A}(V) \to E\widetilde{A}(V) \to B\widetilde{A}(V)$ be the universal fibration. Then the diagram

$$\begin{array}{ccccc}
\widetilde{A}(V)/A(V) & \to & \widetilde{A}(V \times T^m)/A(V \times T^m) & \to & \widetilde{A}(V \times T^n) \times A(V \times T^n) \\
\downarrow & & \downarrow & & \downarrow \\
E\widetilde{A}(V)/A(V) & \to & E\widetilde{A}(V \times T^m)/A(V \times T^m) & \to & E\widetilde{A}(V \times T^n)/A(V \times T^n) \\
\downarrow & & \downarrow & & \downarrow \\
B\widetilde{A}(V) & \to & B\widetilde{A}(V \times T^m) & \to & B\widetilde{A}(V \times T^n)
\end{array}$$

commutes.

Now assume inductively, that $\bar{j}_m \circ \varphi | K^{(r)}$ lifts to $E\widetilde{A}(V \times T^m)/A(V \times T^m)$, $m = (r-1)r/2$. Then the obstruction α to extending this lift to K lies in $H^{r+1}(K; \pi_r(F_m))$, $F_m = \widetilde{A}(V \times T^m)/A(V \times T^m)$. By Corollary 1 the image of α in $H^{r+1}(K; \pi_r(F_n))$, $n = m+r$, is zero. Hence $\bar{j}_n \circ \varphi$ lifts to $E(\widetilde{A}(V \times T^n))/A(V \times T^n)$. But

$$\begin{array}{ccc}
 & \nearrow & BA(V \times T^n) \\
E\widetilde{A}(V \times T^n)/A(V \times T^n) & & \downarrow \\
 & \searrow & B\widetilde{A}(V \times T^n
\end{array}$$

commutes up to homotopy.

By 3.2 and 3.3 we get:

Corollary 3. Assume V is a k-connected PL-manifold of $\dim \geq 5$. Then Corollary 2 holds with $n = r(r+1)/2 - s(s+1)/2$, where

1. $s = \inf(2k-1, k+4)$ with $k \leq n-4$ if ∂V not 1-connected or $\dim V = 5$.
2. $s = \inf(2k, k+4)$, if $k+1 < n/2$, and TV is trivial over the $(k+1)$-skele- skeleton of V.

Corollary 4. Assume K is $(\ell+1)$-connected. Then Corollary 2 holds with $n = r(r+1)/2 - \ell(\ell+1)/2$.

Remarks. Theorem D in the Pl category and for dim K << dim V was first mentioned to us by A. Hatcher. Some form of Corollary 2 we understand is known to Casson, at least in the smooth case.

4. Fibrations over spheres

The aim of this chapter is to apply the results obtained in §3 to the following problem:

Given a compact connected manifold V^{m+d} and a continuous map $f: V^{m+d} \to S^m$, when is f homotopic to a bundle projection $f': V^{m+d} \to S^m$, (i.e., the projection of a locally trivial fibre bundle).

The case $m = 1$ has been completely settled by Browder-Levine and Farell [17]. The case $m = 2$, with the additional restrictions, $\pi_i(f) = 0$, $i \leq 3$, and M smooth and closed, was settled by A. Casson [16].

It is fairly straightforward to generalize the methods of Casson and to combine them with the results in §3 to treat some cases where $m > 2$. Our results are contained in the following theorems 4.1-4.3'.

Theorem 4.1. Let $V^{m+d} \xrightarrow{f} S^m$ be a continuous map from a compact PL manifold V^{m+d} into S^m so that $\partial f = f|\partial V: \partial V \to S^m$ is a PL bundle projection. Let L be the homotopy theoretic fibre of f. Assume $m \geq 2$, $d \geq 5$ and

1) $H_*(L; Z)$ is a finitely generated abelian group;

2) $\pi_i(L) = 0$ if $i \leq s$, where at least one of the following conditions are verified:

   a) $m \leq \inf(2s-1, s+4)$, where $s \leq d-4$ if $(\partial f)^{-1}(x)$ not 1-connected or $d = 5$

   b) $m \leq 2s$, where $s + 1 < d/2$ and $s = 2, 4$ or $s = 1, 3$ and the pull back of τV on the $(s+1)$-skeleton of L is trivial.

Then there is a well-defined obstruction $\nu(f, d) \in P_d$, $P_d = Z, 0, Z_2, 0$, $d \equiv 0, 1, 2, 3 \bmod 4$ respectively, so that $\nu(f, d) = 0$ if and only if f is homotopic rel ∂V to a PL bundle projection.

Corollary 4.2. Let V^{m+d} be a closed PL manifold, $m \geq 2$, $d \geq 5$ and $f: V^{m+d} \to S^m$ a continuous map. Assume the homotopy theoretic fibre L is s-connected and $H_*(L, Z)$ is finitely generated, and that one of the following holds

a) $m \leq \inf(2s-1, s+4)$

b) $m \leq 2s$, where $s+1 < d/2$ and $s = 2, 4$ or $s = 1, 3$ and the pull back of τV on the $(s+1)$-skeleton of L is trivial.

Then f is homotopic to a PL bundle projection if and only if $\nu(f, d) = 0$.

Corollary 4.2 is Theorem 4.1 when $\partial V = \emptyset$. If $\partial V \neq \emptyset$ and we allow f_∂ to move also, the surgery obstruction $\nu(f, d)$ is zero and we get:

Theorem 4.3. Let $V^{m+d} \xrightarrow{f} S^m$ be a continuous map from a compact PL manifold V^{m+d} into S^m with $\partial V \neq \emptyset$ and let L (∂L) be the homotopy theoretic fibre of f (∂f). Assume $m \geq 2$, $d \geq 5$ and

1) $H_*(L, Z)$ and $H_*(\partial L, Z)$ are finitely generated ;

2) $\pi_i(L) = \pi_i(\partial L) = 0$ if $i \leq s$, where at least one of the following conditions are verified:

a) $m \leq \inf(2s-1, s+4)$

b) $m \leq 2s$, where $s+1 < (d-1)/2$ and $s = 2, 4$ or $s = 1, 3$ and the pull back of τV on the $(s+1)$-skeleton of L is trivial.

Then f is homotopic to a PL bundle projection.

If V is smoothable, we get considerably stronger results:

Theorem 4.2'. Let V^{m+d} be a closed PL manifold, $m \geq 2$, $d \geq 5$ and $f: V^{m+d} \to S^m$ a continuous map. Assume V is smoothable, the homotopy theoretic fibre L is s-connected, $H_*(L, Z)$ is finitely generated and that one of the following holds

a) $m \leq 2s-1$

b) $m \leq 2s$, $s+1 < d/2$, and $s = 2, 4, 5, 6 \mod 8$ or , for some smoothing of V the pull back of the tangent vector bundle on the $(s+1)$-skeleton of L is trivial.

Then f is homotopic to a PL bundle projection if and only if $\nu(f, d) = 0$.

Theorem 4.3'. Let $V^{m+d} \xrightarrow{f} S^m$ be a continuous map from a compact PL manifold V^{m+d} into S^m with $\partial V \neq \emptyset$ and let L (∂L) be the homotopy theoretic fibre of f (∂f). Assume V is smoothable, $m \geq 2$, $d \geq 5$ and

1) $H_*(L, Z)$ and $H_*(\partial L, Z)$ are finitely generated;

2) $\pi_i(L) = \pi_i(\partial L) = 0$ if $i \leq s$, where one of the following holds:

a) $m \leq 2s-1$,

b) $m \leq 2s$, $s+1 < (d-1)/2$ and $s = 2, 4, 5, 6 \mod 8$ or the pull back of the tangent bundle on the $(s+1)$-skeleton of L is trivial, for some smoothing of V.

Then f is homotopic to a PL bundle projection.

Proof of Theorem 4.1. The proof will be in three steps:

Step 1. Assuming 1) and 2) we define $\nu(f, d) \in P_d$ and show its vanishing is a necessary and sufficient condition for f to be homotopic rel ∂V to a new f for which the following holds.

a) $f(f^{-1}(S^m_+) \to S^m_+$ is a PL-bundle projection.

b) If $a, b \in S^m_+$, $a \neq b$, then $(f^{-1}(a), \partial f^{-1}(a)) \to (V - f^{-1}(b), \partial V - f^{-1}(b))$ induces an isomorphism for homotopy groups.

Let F be the fibre of $f|f^{-1}(S^m_+)$.

Step 2. Consider pairs (W, g), $g: W \to S^m$ with $g|\partial W$ a PL-bundle projection which satisfies a) and b) with fibre F. Call (W, g) equivalent to (W', g') if there exists a PL-homeomorphism $k: W \to W'$ so that $g = g' \circ k$ on $g^{-1}(S^m_+) \cup \partial W$.

We show that the equivalence classes are in bijective correspondence with $\pi_{m-1}(\tilde{A}^{p\ell}(F,\underset{\sim}{\partial F}))$, where $\tilde{A}^{p\ell}(F,\underset{\sim}{\partial F})$ is the Δ-group whose k-simplices are PL homeomorphisms $\varphi: \Delta_k \times F \to \Delta_k \times F$, $\varphi(d_i\Delta_k \times F) \subset d_i\Delta_k \times F$, and $\varphi|\Delta_k \times \partial F$ commutes with projection into Δ_k.

Step 3. We check that the results of §3 imply that $\pi_{m-1}(A^{p\ell}(F,\partial F)) \to \pi_{m-1}(\tilde{A}^{p\ell}(F,\underset{\sim}{\partial F}))$ is surjective under either of the hypotheses a) or b), where $A^{p\ell}(F,\partial F)$ is the Δ-group of PL-bundle homeomorphisms $\varphi: \Delta_k \times F \to \Delta_k \times F$ such that φ commutes with projection on Δ_k (i. e., $\varphi|\Delta_k \times \partial F$ is not assumed the identity). If $\alpha \in \pi_{m-1}(\tilde{A}^{p\ell}(F,\underset{\sim}{\partial F}))$ is in the image of $\pi_{m-1}(A^{p\ell}(F,\partial F))$ we show there exists $(W,g) \in \alpha$ which is a PL-bundle projection. Thus the f obtained in Step 1 when $\nu(f,d) = 0$, is homotopic to a PL-bundle projection.

Step 1 is essentially Casson's theorem [16], but we prefer to sketch the proof by indicating more appropriate references since we work in the PL category and to take care of the case $\partial V \neq \emptyset$.

Step 1. Let $f: V \to S^m$ be a map such that $f: \partial V \to S^m$ is a locally trivial PL-bundle. Let $\emptyset: X \to S^m$ be the associated Hurewicz fibre map, i. e., $X = \{(v,\lambda),\ v \in V,\ \lambda: I \to S^m \text{ such that } \lambda(0) = f(v)\}$, $\emptyset(v,\lambda) = \lambda(1)$. Then $i: V \to X$, $i(v) = (v, \text{constant})$, is a homotopy equivalence and $f = \emptyset \circ i$. The fibre L of $\emptyset$ is the homotopy theoretic fibre of f. We define $\partial X = i(\partial V)$ and $\partial L = \partial X \cap L$. Suppose now $b \in S^m$ is a regular point for f; i. e., $f|f^{-1}(U_b) \to U_b$ is a locally trivial bundle, U_b an open neighborhood of b in S^m. Then $i: (f^{-1}(b), \partial f^{-1}(b)) \to (L, \partial L)$ is a map which restricted to $\partial f^{-1}(b)$ is a homeomorphism.

Now $(X, \partial X)$ is a Poincaré duality pair of dim m+d, $m \geq 2$. Since $H_n(L) = 0$ for all sufficiently large n, it is easy to check that $(L, \partial L)$ is also a Poincaré duality pair. Further, if $b \in S^m$ is a regular value of f,

$i: (f^{-1}(b), \partial f^{-1}(b)) \to (L, \partial L)$ has degree 1. The proofs go almost exactly as in Casson [16].

Let ξ be a normal microbundle for V, and $\xi' = (i^{-1})^{*}\xi$ the induced microbundle over X. We may assume $U_b \supset S^m_+$. Then for $a \in S^m_+$ we let $\varepsilon = \xi | f^{-1}(a)$ and $\varepsilon' = \xi' | L$. Then we have $i: (f^{-1}(a), \partial f^{-1}(a)) \to (L, \partial L)$, i a homeomorphism on $\partial f^{-1}(a)$; and i is covered by a microbundle map $i_*: \varepsilon \to \varepsilon'$. Further ε is the normal bundle of $f^{-1}(a)$, since $f^{-1}(a)$ has a trivial normal bundle in V.

We define $\nu(f, d)$ to be the surgery obstruction to making (i, i_*) normally cobordant rel boundary to (j, j_*) with j a homotopy equivalence of pairs. We note that $\nu(f, d)$ does not depend on a, and only depends on f up to a homotopy equivalence rel boundary, because $\nu(f, d)$ is a normal cobordism invariant. $\nu(f, d)$ lies in P_d because $\pi_1(L) = 0$. Clearly, f homotopic to a fibration implies $\nu(f, d) = 0$. Actually, if f satisfies a) and b), it is not difficult to see that i is a homotopy equivalence of pairs and hence that $\nu(f, d) = 0$ (see Step 2 below or [16]).

It remains to show that $\nu(f, d) = 0$ implies that f is homotopic to f' satisfying a) and b). Now if $\nu(f, d) = 0$, we have a normal cobordism $(\tilde{i}, \tilde{i}_*)$:

$$\begin{array}{ccc} \tilde{\varepsilon} & \xrightarrow{\;i_*\;} & \varepsilon' \times I \\ \downarrow & & \downarrow \\ W & \xrightarrow{\;i_*\;} & L \times I \end{array}$$

with $\partial W = \partial f^{-1}(a) \times I$, $\partial_0 W = f^{-1}(a)$ and $\tilde{i} | \partial W = i \times \mathrm{id}$, $\tilde{i} | \partial_0 W = i$ and $\tilde{i} | \partial_1 W = j$ a homotopy equivalence of pairs. Let $F = \partial_1 W$. Then $\partial F = f^{-1}(a)$.

Now follow the argument of Theorem 11.3 (relative) in Wall [18]. Construct a Poincare embedding of F in V with trivial normal bundle by choosing a fibre homotopy equivalence

$$t: F \times S^m_- \cup_h F \times S^m_+ \to X \quad \text{and a homotopy}$$

$$\begin{array}{ccc} F \times S^m_- \cup_h F \times S^m_+ & \longrightarrow & X \\ & \searrow \quad \swarrow & \\ & S^m & \end{array}$$

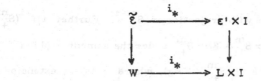

equivalence $h: F \times S^{m-1} \to F \times S^{m-1}$ such that the restrictions

$$\begin{array}{ccc} F \times S^{m-1} & \xrightarrow{h} & F \times S^{m-1} \\ & \searrow \quad \swarrow & \\ & S^{m-1} & \end{array}$$

$h: \partial F \times S^{m-1} \to \partial F \times S^{m-1}$, and $t: \partial F \times S^m_- \cup_h \partial F \times S^m_+ \to \partial X$ are PL-homeomorphisms. Let $s: (X, \partial X) \to (V, \partial V)$ be a homotopy inverse (rel boundary) of i with $s|\partial X = i^{-1}$. Then $st: F \times S^m_- \cup_h F \times S^m_+ \to V$ is a Poincaré embedding satisfying the conditions of Theorem 11.3 of [18].

Let μ be a homotopy inverse to st. Now the cobordism W of $f^{-1}(a)$ extends to a cobordism of $(\mu, V, f^{-1}(a))$ which we can assume (after surgery (see 11.5 of [18])) is of the form $(V \times I, W)$, with W having a trivial normal bundle $W \times S^m_+$ in $V \times I$. Further, we get a homotopy equivalence $\rho: V \times 1 \to F \times S^m_- \cup_h F \times S^m_+$ which is the identity on $F \times S^m_+$ and a homotopy equivalence of the closure of the complements, and with $\rho|\partial V \times 1$ the homeomorphism $(st)^{-1}$.

By the Pontrjagin construction, i.e., collapsing everything outside of the normal tube of W in $V \times I$ to a point, we see that f is homotopic to f' satisfying a) with fibre F. f' also satisfies b) by the properties of ρ.

Step 2. First note that a) and b) of Step 1 imply that V is PL-homoemorphic to $S^m_- \times F \cup_h S^m_+ \times F$ where $h: S^{m-1} \times F \to S^{m-1} \times F$ is a PL-homeomorphism such that $h|S^{m-1} \times \partial F$ commutes with projection in S^{m-1}. Further $f|f^{-1}(S^m_+)$ corresponds to the projection $p: S^m_+ \times F \to S^m_+$ under the homeomorphism. Thus (V, f) is equivalent to $(S^m_- \times F \cup_h S^m_+ \times F; g)$, where g is any extension of ph^{-1} to a map sending $S^m_- \times F \to S^m_-$. Now h represents an element of $\pi_{m-1}(A^{P\ell}(F, \partial F))$ and any (V', f') equivalent to (V, f) leads to an h' representing the same homotopy class.

Conversely, $\alpha \in \pi_{m-1}(A^{P\ell}(F, \partial F))$ is represented by such an h, and $(S^m_- \times F \cup_h S^m_+ \times F; g)$ is a pair (W, g) well-defined up to equivalence

Step 3. Consider

$$\begin{array}{ccccc} A^{p\ell}(F) & \to & A^{p\ell}(F,\partial F) & \to & A^{p\ell}(\partial F) \\ \downarrow & & \downarrow & & \downarrow \\ \widetilde{A}^{p\ell}(F) & \to & \widetilde{A}^{p\ell}(F,\partial F) & \to & A^{p\ell}(\partial F) \end{array}$$

where the horizontal lines are fibrations. Then $\pi_{m-1}(A^{p\ell}(F,\partial F)) \to \pi_{m-1}(\widetilde{A}^{p\ell}(F,\partial F))$ will be surjective if $\pi_i(A^{p\ell}(F)) \to \pi_i(\widetilde{A}^{p\ell}(F))$ is bijective for $i \leq m-2$ and surjective for $i = m-1$. This in turn is implied by $\pi_i^{rel}(A^{p\ell}(F)) = 0$ for $i \leq m-1$. Under any of the conditions a) or b) this last follows from (3.2) and (3.3) (see Remark following Corollary 3.3).

Remark. As we noted $\nu(f,d)$ is always zero for d odd. For $d = 4k$, $\nu(f,d)$ may be computed as follows:

$$\nu(f,d) = \frac{1}{8} < L_k(V) \cup f^*[S^m]; [V] > ,$$

where $[S^m], [V]$ denote the fundamental cycles of S^m and $(V, \partial V)$ respectively, and L_k is the Hirzebruch polynomial (see Casson [16]).

Proof of Theorem 4.3. We show how to modify the proof of 4.1 to fit this case.

Step 1. Take $\emptyset : X \to S^m$ as in 4.1. Define $\partial X = \{(v,\lambda), v \in \partial V, \lambda : I \to S^m$, such that $\lambda(0) = f(v)\}$. Define ∂L = fibre of $\emptyset | \partial X$. Then $i: (V, \partial V) \to (X, \partial X)$ is a homotopy equivalence of pairs and $(L, \partial L)$ is a Poincaré duality pair. Now in this case (i, i_*), $i: (f^{-1}(a), \partial f^{-1}(a)) \to (L, \partial L)$, is always normally cobordant to (j, j_*) with j a homotopy equivalence of pairs, since we do not fix the boundary. The rest of Step 1 proceeds as before.

Step 2. Again conditions a) and b) of Step 1 imply V is homeomorphic to $S^m_- \times F \cup_h S^m_+ \times F$. Only now, h represents an element of $\pi_{m-1}(\widetilde{A}^{p\ell}(F, \partial F))$, where a k-simplex of $\widetilde{A}^{p\ell}(F, \partial F)$ is a PL-homeomorphism $\varphi: \Delta_k \times F \to \Delta_k \times F$, $\varphi(d_i\Delta_k \times F) \subset d_i\Delta_k \times F$; we do not assume $\varphi | \Delta_k \times \partial F$ commutes with projection.

Step 3. Consider

$$\begin{array}{ccccc} A^{p\ell}(F) & \to & A^{p\ell}(F,\partial F) & \to & A^{p\ell}(\partial F)\\ \downarrow & & \downarrow & & \downarrow \\ \tilde{A}^{p\ell}(F) & \to & \tilde{A}^{p\ell}(F,\partial F) & \to & \tilde{A}^{p\ell}(\partial F)\end{array}$$

Again $\pi_{m-1}(A^{p\ell}(F,\partial F)) \to \pi_{m-1}(\tilde{A}^{p\ell}(F,\partial F))$ will be surjective if $\pi_i(A^{p\ell}(F)) \to \pi_i(\tilde{A}^{p\ell}(F))$ is surjective for $i = m-1$ and bijective for $i = m-2$, and also $\pi_{m-1}(A^{p\ell}(\partial F)) \to \pi_{m-1}(\tilde{A}^{p\ell}(\partial F))$ is surjective. This in turn is implied by $\pi_i^{rel}(A^{p\ell}(\partial F)) = 0$ for $i \leq m-1$. Under any of the conditions a), b) this last follows from (3. 2) or (3. 3).

Proof of Theorem 4. 2'. We apply the argument of 4. 1 in the smooth category; i. e., Casson's argument. Since F is a closed smooth manifold, we replace $\tilde{A}^{p\ell}(F,\partial F)$ by $\tilde{A}^d(F)$ in Step 2. Then $f: V \to S^m$ will be homotopic to a PL-bundle projection if we can show $\pi_{m-1}(\tilde{A}^d(F)) \to \pi_{m-1}(\tilde{A}^{p\ell}(F))$ lifts to $\pi_{m-1}(A^{p\ell}(F))$. But this follows if $\pi_i^{rel}(A^d(D)) \to \pi_i^{rel}(A^d(F))$ is bijective for $i < m-1$ and surjective for $i = m-1$, where D is a disc of codimension zero in F, since this is equivalent to $\pi_i(A^d(F), A^d(D)) \to \pi_i(\tilde{A}^d(F), \tilde{A}^d(D))$ bijective for $i < m-1$ and surjective for $i = m-1$ which in turn implies the result by the following commutative diagram:

$$\begin{array}{ccccc} \pi_{m-1}(A^d(F), A^d(D)) & \to & \pi_{m-1}(A^{p\ell}(F), A^{p\ell}(D)) & \cong & \pi_{m-1}(A^{p\ell}(F)) \\ \downarrow & & \downarrow & & \downarrow \\ \pi_{m-1}(\tilde{A}^d(F), \tilde{A}^d(D)) & \to & \pi_{m-1}(\tilde{A}^{p\ell}(F), A^{p\ell}(D)) & \cong & \pi_{m-1}(\tilde{A}^{p\ell}(F) \\ \uparrow & & & \nearrow & \\ \pi_{m-1}(\tilde{A}^d(F)) & & & & \end{array}$$

Any of the conditions a), b) imply the above by (3. 2 and (3. 3).

Proof of Theorem 4.3'. We apply the argument of 4.3 in the smooth category. In Step 2 we replace $\tilde{A}^{p\ell}(F, \partial F)$ by $\tilde{A}^d(F, \partial F)$, the corresponding smooth complex (see Chapter 2). Let $D^d_+ \subset F$ be a smooth embedding with $\partial F \cap D^d_+ = D^{d-1}$, $\partial D^d_+ - \mathrm{Int}\, D^{d-1}$ transverse to ∂F. Let $\tilde{A}^d(D^d_+, D^{d-1})$ be the complex whose k-simplices are diffeomorphisms $\varphi: \Delta_k \times D^d_+ \to \Delta_k \times D^d_+$, $\varphi(d_i\Delta_k \times D^d_+) \subset d_i\Delta_k \times D^d_+$ and $\varphi | \Delta_k \times S^d_+ =$ identity. Consider the map $\tilde{A}^d(D^d_+, D^{d-1}) \to \tilde{A}^d(F, \partial F)$ obtained by extending φ by the identity outside D^d_+. Then we have a map of fibrations

$$\begin{array}{ccccc} \tilde{A}^d(D^d_+) & \to & \tilde{A}^d(D^d_+, D^{d-1}) & \to & \tilde{A}^d(D^{d-1}) \\ \downarrow & & \downarrow & & \downarrow \\ \tilde{A}^d(F) & \to & \tilde{A}^d(F, \partial F) & \to & \tilde{A}^d(\partial F) \end{array}$$

The simplices of $\tilde{A}^d(D^d_+)$ are the identity on $\Delta_k \times \partial D^d_+$, and the subcomplex of those that are the identity on a neighborhood of $\Delta_k \times \partial D^d_+$ is a deformation retract. Removing a collar neighborhood of D^d_+ we get a disc $D^d \subset \mathrm{Int}\, F$. Hence we have

$$\begin{array}{ccc} \tilde{A}^d(D^d) & \simeq & \tilde{A}^d(D^d_+) \\ & \searrow \swarrow & \\ & \tilde{A}(F) & \end{array}$$

commutes. We there fore get an exact sequence:

$$\to \pi_i(\tilde{A}^d(F), \tilde{A}(D^d)) \to \pi_i(\tilde{A}^d(F, \partial F)), \tilde{A}^d(D^d_+, D^{d-1})) \to \pi_i(\tilde{A}^d(\partial F), \tilde{A}^d(D^{d-1})) \to$$
$$\pi_{i-1}(\tilde{A}^d(F), \tilde{A}^d(D^d)) \to \cdots$$

and a corresponding exact sequence (without the tildas) for the subcomplex of φ's commuting with projection, which maps into the above sequence. Thus if $\pi_i^{rel}(A^d(D^d)) \to \pi_i^{rel}(A^d(F))$ is bijective for $i < m-1$ the same holds for $\pi_i(A^d(F), A^d(D^d)) \to \pi_i(\tilde{A}^d(F), \tilde{A}^d(D^d))$, and likewise $\pi_i^{rel}(A^d(D^{d-1})) \to \pi_i^{rel}(A^d(\partial F))$ and hence $\pi_i(A^d(\partial F), A^d(D^{d-1})) \to \pi_i(\tilde{A}^d(\partial F), \tilde{A}^{d-1}(D^{d-1}))$, and we get

$\pi_i(A^d(F, \partial F), A^d(D^d_+, D^{d-1})) \to \pi_i(\tilde{A}^d(F, \partial F), \tilde{A}^d(D^d_+, D^{d-1}))$ is bijective for $i < m-1$ and surjective for $i = m-1$.

Now $A^{p\ell}(D^d_+, D^{d-1})$ and $\tilde{A}^{p\ell}(D^d_+, D^{d-1})$ analogously defined, are contractible by the Alexander trick. Hence a similar diagram (with $(F, \partial F)$ in place of F, and (D^d_+, D^{d-1}) in place of D^d) to that in the proof of 4. 2' shows that $\pi_{m-1}(\tilde{A}^d(F, \partial F)) \to \pi_{m-1}(\tilde{A}^{p\ell}(F, \partial F))$ lifts to $\pi_{m-1}(A^{p\ell}(F, \partial F))$, and hence f is homotopic to a PL-bundle map if one of the conditions a) or b) holds.

5. Fibrations over Manifolds

Section 0.

In the last chapter, we examined the problem of deforming $f: V \to S^m$ to a bundle projection. We now consider this problem when S^m is replaced by an arbitrary manifold. We work simultaneously in the $p\ell$ and smooth category. The machinery of this chapter is somewhat more formidable than that assumed in earlier chapters. We assume familiarity with the ideas of Sections 0-12 of Surgery on Compact Manifolds of C. T. C. Wall [18], although our notation is slightly different from his. One should also see Section 17A of that book, in particular the references to Quinn [22] whose work anticipates many of the constructions here. Our main results are Theorems 5.0.1 and its corollaries along with the results of Section 5 and the reader might wish to examine them before going through the entire text.

We fix $f: V^{m+d} \to W^m$. The following picture should be kept in mind.

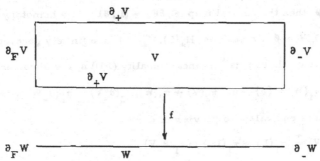

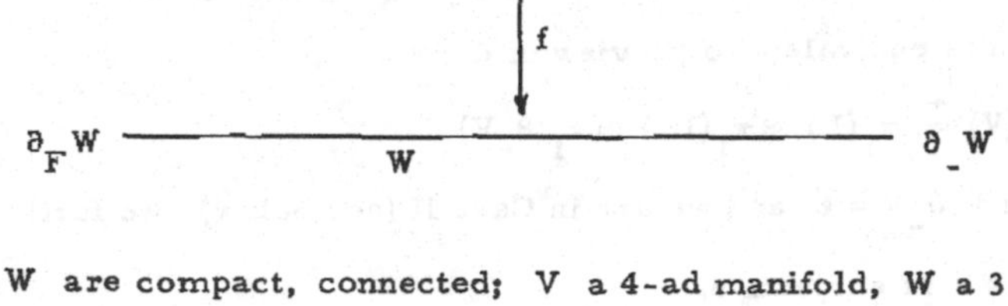

Explicitly: V, W are compact, connected; V a 4-ad manifold, W a 3-ad manifold, $\partial V = \partial_F V \cup \partial_- V \cup \partial_+ V$, $\partial W = \partial_F W \cup \partial_- W$, $f^{-1}(\partial_F W) = \partial_F V$, $f^{-1}(\partial_- W) = \partial_- V$. That is, f is a 3-ad map.

We make the following set of assumptions.

a) $V, W, \partial_- V$ are connected;

b) $d \geq 5$ and ($d \geq 6$ if $\partial_+ V \neq 0$ and we are in Case II (see below)),

c) $f | \partial_F V \to \partial_F W$ is the projection of a locally trivial fibre bundle.

Let L = homotopy theoretic fibre of f,

L_- = homotopy theoretic fibre of $f_- = f | \partial_- V \to \partial_- W$,

$\partial L = L_+$ = homotopy theoretic fibre of $f_+ = f | \partial_+ V \to W$,

∂L_- = homotopy theoretic fiber of $f_- = f | \partial_+ V \cap \partial_- V$,

F = fiber of $f | \partial_F V \to \partial_F W$.

$L, L_-, \partial L, \partial L_-$ are determined up to homotopy type while F is determined up to (smooth or $p\ell$) isomorphism over each component of $\partial_F W$.

We assume

d) if $\partial_- V \neq \emptyset$ then the natural map $(L_-, \partial L_-) \to (L, \partial L)$ is a homotopy equivalence.*

e) if $\partial_F W \neq \emptyset$ then the natural map $(F, \partial F) \to (L, \partial L)$ is a homotopy equivalence of pairs. If $\partial_F W = \emptyset$ we assume $H_*(L), H_*(\partial L)$ are finitely generated. This implies [22] that $(L, \partial L)$ is a Poincaré duality (PD) space of formal dimension d.

f) $\pi_1(\partial_- W) = \pi_1(W) = \{e\}$ and $\pi_2(V) \to \pi_2(W), \pi_2(\partial_- V) \to \pi_2(\partial_- W)$ are surjective. This condition is equivalent to (in view of d)

$$\pi_1(V) \overset{\leftarrow}{\approx} \pi_1(L) \approx \pi_1(L_-) \approx \pi_1(\partial_- V).$$

If $\partial_+ V \neq \emptyset$ and $\partial_- V = \emptyset$ and we are in Case II (see below), we further assume $\pi_2(\partial_+ V) \to \pi_2(W)$ is surjective.

---

*By homotopy equivalence we will mean weak homotopy equivalence, i. e., the map induces an isomorphism on components and homotopy groups.

We will say we are in <u>Case I</u> if $f|\partial_+ V$ is the projection of a fiber bundle which is to remain constant during our deformations. Otherwise we say we are in <u>Case II.</u>

The questions we then ask are:

Case I: When is f homotopic rel $\partial_F V \cup \partial_+ V$ to a bundle projection?

Case II: When is f homotopic rel $\partial_F V$ to a bundle projection?

A careful analysis of the spherical case indicates the problem decomposes into two problems. The first is to deform f to a special form, called a block bundle projection. The second is to deform a block bundle projection to a bundle projection. We will devote the next few sections to the first of these problems.

Let us fix once and for all a triangulation of W as a $p\ell$ manifold with $\partial_- W, \partial_F W$ as $p\ell$ submanifolds. In the smooth category, we use smooth triangulations.

Our assumptions guarantee that L is a PD space. For f to be homotopic to a bundle projection, however, we must have more. Namely, that $(L, \partial L)$ is homotopically equivalent to a finite (or finite simple) PD space [18]. If $\partial_F V \neq \emptyset$, this is guaranteed by (e) since $(F, \partial F)$ is a finite PD space. In general, if $\partial_F V = \emptyset$, there is an obstruction which depends on $\pi_1(L), \pi_1(\partial L)$ and the map $\pi_1(\partial L) \to \pi_1(L)$ [23]. In particular, if $\pi_1(\partial L) \cong \pi_1(L)$, the obstruction vanishes.

More generally, let K be a topological n-ad. An <u>s-structure on</u> K is determined by a homotopy equivalence on n-ads $h: C \to K$ where C is a finite CW n-ad. h_1 and h_2 determine the <u>same</u> structure if there exists $r: C_1 \to C_2$ with $h_2 r \sim h_1$, where r is a <u>simple</u> homotopy equivalence ($\sim$ means homotopic). For n-ads with an s-structure, the notion of simple homotopy equivalence and simple homotopy type is defined. We also refer to a simple homotopy equivalence as an s-equivalence and denote it by $\overset{s}{\sim}$.

We now make assumption g .

g) If $\partial_F V \neq \emptyset$ then the s-structure on $(L, \partial L)$ induced from distinct components of $\partial_F V$ agree. If $\partial_F V = \emptyset$ we assume we can and have selected an s-structure on $(L, \partial L)$ which makes it a finite PD pair in the sense of [18]. If we are in Case I we assume the structure agrees with the one on ∂L already determined by $f^{-1}(p) \cap \partial_+ V \xrightarrow{\simeq} \partial L$.

Proposition 5.0. Let $\pi: E \to K$ be a Serre fibration K-connected, simply connected and a finite CW complex. Let $p \in K$ and $h_p: F \to \pi^{-1}(p)$ be an s-structure. Then there exists a unique s-structure $h = h_K: C \to E$ on E such that (a) for any subcomplex $K' \subset K$, $h_{K'}: C_{K'} \to \pi^{-1}(K')$ is an s-structure, where $C_{K'} = h^{-1}\pi^{-1}(K')$ is a subcomplex of C, $h_{K'} = h|C_{K'}$, (b) h_p is the h_p described above and thus $C_p = F$. Further, if K is a simplicial complex and $\Delta_k \subset K$ is a k-simplex, then we have $C_{\Delta_k} \overset{s}{\simeq} F \times \Delta_k$ as (k+2)-ads.

The proof is a simple pasting argument using properties of fibrations and elementary properties of simple homotopy equivalences [24]. (See [25, Corollary 3.16] for a more elegant formulation of this result, although they don't deal explicitly with the notion of simple homotopy equivalence.)

The space C constructed above is called a block fibration with fiber F and base K [25]. It leads to the following notion.

If X is an $(n+r_1)$-ad space, Y an $(n+r_2)$-ad space, an n-ad map $f: X \to Y$ is a map such that $f^{-1}(Y(i)) = X(i)$, $0 \leq i \leq n-2$. If X and Y are manifolds and if Y is triangulated with each face $Y(\alpha)$ a subcomplex, we say f is transverse if f is transverse to each simplex of Y, $f(i): X(i) \to Y(i)$ is transverse to each simplex of $Y(i)$, $0 \leq i \leq n-2$, and $f|X(j): X(j) \to Y$ is transverse to each simplex of Y for all j.

We say that f is a <u>block bundle projection</u> if Y is an n-ad and if f is a transverse n-ad map such that for each pair of simplices $\Delta_i, c\Delta_j, cY$, then

$$(f^{-1}(\Delta_i), f_+^{-1}(\Delta_i)) \to (f^{-1}(\Delta_j), f_+^{-1}(\Delta_j))$$

is a simple homotopy equivalence of pairs.

In our situation with $X = V$, $Y = W$, we further assume the natural map $(f^{-1}(p), f_+^{-1}(p)) \to (L, \partial L)$ is an s-equivalence for one and hence every vertex p of W.

If $h: C \to E$ is an s-structure as given in 5.0 on $\pi: E \to K$, with πh transverse, then πh is a block bundle projection.

We now fix once and for all $\pi: E \to W$ to be the Hurewicz fibration associated to f and in Case II, $\pi_+: E_+ \to W$ the Hurewicz fibration associated with $f_+ = f|\partial_+ V$ while in Case I, $E_+ = \partial_+ V$ and $\pi_+ = \pi|E_+$, where $\partial_+ V$ is naturally embedded in E. Then in both cases, we can consider $E_+ \subset E$, $\pi_+ = \pi|E_+$. We let $E_F = \pi^{-1}(\partial_F W)$ and $E_- = \pi^{-1}(\partial_- W)$. We then have an induced map of 4-ads:

$$r: (V, \partial_F V, \partial_- V, \partial_+ V) \to (E, E_F, E_-, E_+),$$

with $\pi r = f$. The assumptions (d, e) guarantee r is a homotopy equivalence (weak) of 4-ads. Proposition 5.0 and (g) provide an s-structure on E.

By the uniqueness part of 5.0 and the covering homotopy property of π, if f is homotopic to a bundle projection r must be an s-equivalence. In fact, if f is homotopic to a block bundle projection r is an s-equivalence. Therefore we must add assumption

h) r is an s-equivalence.

<u>Remark.</u> This assumption is the most unpleasant one since it is the most difficult in practice to verify. If $\partial_F V = \emptyset$ we can attempt to realize it by altering the s-structure of $(L, \partial L)$, but the effect of altering the s-structure of $(L, \partial L)$ on the

s-structure of (E, E_+) doesn't seem to be well understood. At the end of Section 2 we will outline a strategy for evading this problem, at least partially. Note that if all fundamental groups are free abelian or sums of $Z/2Z$, conditions g) and h) are automatically satisfied.

We are now ready to state our first main theorem. For this it is more natural to relax hypothesis c) to the weaker

c') $f|\partial_F V \to \partial_F W$ is a block bundle projection.

Since all deformations of f we use are modulo $\partial_F V$, we do not lose any information in replacing c) by c'), and we assume we have done so.

Theorem 5.0.1. There exist a Δ-set $A(f)$, a Kan Fibration $\rho: A(f) \to W$ of Δ-sets, and a crossection γ over $\partial_F W$ such that f is homotopic modulo $\partial_F V \cup \partial_+ V$ in Case I and $\partial_F V$ in Case II to a block bundle projection if and only if γ extends to W.

Further, let $\pi_1 = \pi_1(L) = \pi_1(V)$ be the fundamental groupoid and $w: \pi_1 \to Z_2$ the Stiefel Whitney class. Let $\pi_1^+ = \pi_1(\partial L) = \pi_1(\partial_+ V)$. Then in Case I, $\pi_c(\rho) = L_{c+d}(\pi_1, w)$ and in Case II, $\pi_c(\rho) = L_{c+d}(\iota)$, where $\iota: (\pi_1^+, w^+) \to (\pi_1, w)$ is induced by $\partial L \to L$ and the $L_{c+d}(\ \ )$ groups are the surgery obstruction groups of Wall [18].

Corollary 5.0.2. If $\iota: \pi_1^+ \to \pi_1$ is an isomorphism then f is homotopic to a block bundle projection in Case II -- since $L^k(\iota) = (e)$ when ι is an isomorphism.

Let K be any connected manifold. Let f_k be the composition $V \times K \to V \xrightarrow{f} W$. Let $V_K = V \times K$, $\partial_F V_K = \partial_F V \times K$, $\partial_- V_K = \partial_- V \times K$, $\partial_+ V_K = \partial_+ V \times K \cup V \times \partial K$. Then $f_K: V_K \to W$ also satisfies hypotheses a)-h).

For the following we assume we are in Case II.

Corollary 5.0.3. Let K be any connected manifold with connected boundary and $\pi_1(\partial K) = \pi_1(K)$. Then $f_K: V \times K \to W$ is homotopic to a block bundle projection because in this case, $\pi_1(\partial_+ V_K) \simeq \pi_1(V_K)$.

Corollary 5.0.4. Suppose K is closed, $K = \partial K'$, K is connected, and $\pi_1(K) = \pi_1(K')$. Suppose $\partial_F V = \emptyset$. Then $f_K: V \times K \to M$ is homotopic to a block bundle projection because $V \times K = \partial_+(V_{K'})$ and $f_K = f_{K'}$.

Remark. If $f: X \to Y$ is a homotopy equivalence, then $f \times \mathrm{id}: X \times X^1 \to X \times S^1$ is an s-equivalence. Thus if $f: V \to W$ satisfies all assumptions except h), then $f_{S^1}: S^1 \times V \to W$ satisfies all the assumptions.

Section I.

In this section we begin our proof of Theorem 5.0.1. We show that it is equivalent to a cobordism problem. We begin with some definitions which we use throughout.

Let X be an $(n+2)$-ad. Recall that X has $n+2$ faces of codimension 1 (which we denote henceforth by $X(0),\ldots,X(n)$). Thus an n-simplex Δ_n is an $(n+2)$-ad with $\Delta_n(i) = \partial_i\Delta_n$. Each $X(i)$ is in turn an $(n+1)$-ad. Let δX be the $(n+3)$-ad with the same ambient space, $\delta X = X$, $\delta X(i) = X(i)$, $0 \le i \le n$ and $\delta X(n+1) = \emptyset$. A cobordism of X^1 to X^2 is an $(n+4)$-ad C and two isomorphisms $\delta_1 : \delta X^1 \approx C(n+1)$, $\delta_2 : \delta X^2 \approx C(n+2)$ of $(n+3)$-ads. The canonical example is $X^1 = X^2 = X$ and $C = X \times I$. (The product of an $(n+2)$-ad X and a 3-ad I is an $n+2+3-1 = (n+4)$-ad.) Note that this definition requires $\delta_1(\delta X^1) \cap \delta_2(\delta X^2) = \emptyset$. Ordinarily, to avoid proliferation of notation we suppress δ_1, δ_2, identify δX with X and consider X^1, X^2 as subsets of C. However, this is clearly illegitimate in the case when, for example, $X^1 = X^2$. With this warning we will utilize the simpler notation.

Observe that with this convention, for $0 \le j \le n$, $C(j)$ is a cobordism of $X^1(j) = C(j) \cap C(n+1)$ and $X^2(j) = C(j) \cap C(n+2)$, considered now as $(n+1)$-ads. If $C(j_t) = I \times X^1(j_t)$ $(0 \le t \le k)$ we say C is a cobordism modulo $(j_0,\ldots,j_k)$. For example, $I \times X$ is a cobordism modulo $(0,\ldots,n)$. A cobordism modulo $(0,\ldots,n)$ will be called a cobordism modulo ∂X.

If X^1, X^2 are manifold $(n+2)$-ads, we always assume C is a manifold $(n+4)$-ad.

Recall that for a $(n+2)$-ad X, and $\alpha \subset (0,\ldots,n)$, the α face $X(\alpha)$ of X is defined and is an $(n+2-k)$-ad. In fact, what we have denoted by $X(c)$ before $(0 \le c \le n)$ is in this notation $X((0,\ldots,\hat{c},\ldots,n))$. If C is a cobordism of $(n+2)$-ad manifolds X^1, X^2, we say that C is an s-cobordism if for each $\alpha \subset (0, \ldots, n)$,

the map $X^i(\alpha) \to C(\alpha \cup (n+1, n+2))$ is a simple homotopy equivalence (s-equivalence).

The s-cobordism theorem asserts that under certain dimensionality conditions an

s-cobordism is isomorphic as an $(n+2)$-ad manifold to the product cobordism.

Let V be an $(n+2)$-ad manifold, Y an $(n+2)$-ad PD space, ε a stable vector

bundle (linear or pℓ depending on the category) over Y. We consider a normal

map of degree 1, $\lambda : V \to (Y, \varepsilon)$ (see [18] Ch.3 for definitions). We also use

$\lambda : V \to Y$ (ambiguously) to denote the associated map of $(n+2)$-ads.

A normal cobordism $\overline{\lambda}$ of λ is an $(n+4)$ad normal map of degree 1

$\overline{\lambda}: C \to (I \times Y, I \times \varepsilon)$ where C is a cobordism of V and $\overline{\lambda}(n+1) = \overline{\lambda} \, | \, C(n+1) = \lambda$.

$\overline{\lambda}$ is a normal cobordism modulo $(j_0, \ldots, j_k)$ if C is and $\overline{\lambda}(j_i) = \text{id} \times \lambda (j_i)$.

We now consider our map from the previous section:

$$r: (V, \partial_F V, \partial_- V, \partial_+ V) \to (E, E_F, E_-, E_+) \ .$$

Since r is an s-equivalence, there is a unique stable bundle ε over E with

$r^*(\varepsilon)$ a normal bundle of V. r thus can be extended (not uniquely) to a normal

map of degree 1, still denoted by $r: V \to (E, \varepsilon)$. We now consider normal cobord-

isms of r which we always assume modulo (0), and in case I modulo $(0, 2)$. We

can now state the main result of this section.

Theorem 5.1.0. Under our assumptions a -h , f is homotopic modulo

$\partial_F V$, and in Case I modulo $\partial_F V \cup \partial_+ V$, to a block bundle projection if and only if

r is normally cobordant to one. Explicitly, the condition holds if and only if there

exists a normal cobordism $\overline{r}: C \to (I \times E, I \times \varepsilon)$ of r with $\pi \overline{r}(4): C(4) \to W$ a block

bundle projection.

The remainder of this section will be devoted to a proof of this theorem.

This theorem brings into play our assumption f). It would be interesting to attempt

to weaken that hypothesis.

Proof of Theorem 5.1.0. If f is homotopic to a block bundle projection, we can let $C = I \times V$ and use the covering homotopy property of fibrations to get $\overline{r}: I \times V \to I \times E$ covering the homotopy of f.

To prove the converse, we will alter (r, C) to an s-cobordism and then use the s-cobordism theorem. This alteration will proceed via surgery; however surgery by itself is not sufficient since certain obstructions can occur. At that point we utilize the theory of surgery obstructions as developed in [18] to cancel out the obstructions, relying for this on our control of the various fundamental groups involved, which is guaranteed by assumption f).

To be more explicit we consider the following two types of alterations of (r, C).

0. Relative surgery. (See [18].

We can alter $(\overline{r}, C)$ by doing relative surgery on $\overline{r}: (C, \partial C) \to (I \times E, \partial(I \times E), I \times \varepsilon)$. This surgery is always modulo $C(0) \cup C(3) \cup C(4)$, and in case I, also modulo $C(2) = I \times \partial_+ V$. These surgeries are to respect the 6-ad structure of $(\overline{r}, C)$.

1. Interior pasting.

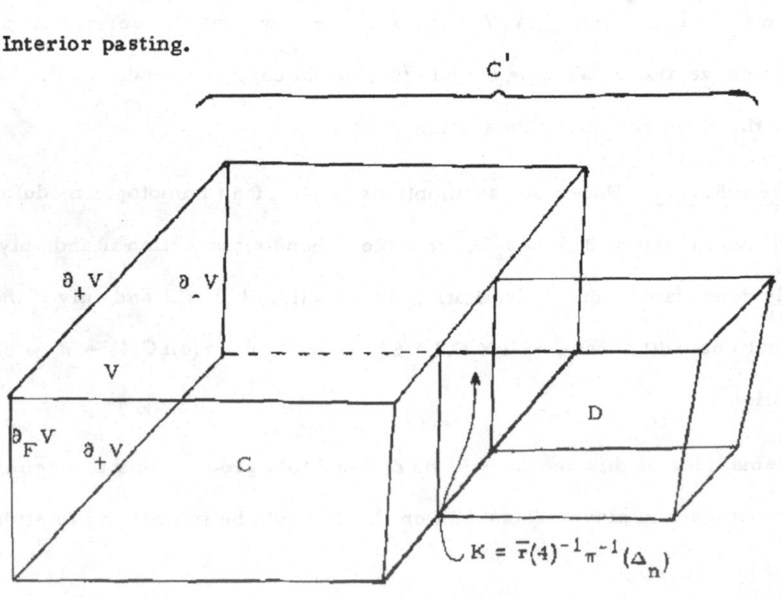

Let Δ_n be a simplex of W of maximal dimension with $\Delta_n \cap \partial W = \emptyset$. We are going to alter $(\bar{r}, C)$ by pasting to C a cobordism D of $\bar{r}(4)^{-1}\pi^{-1}(\Delta_n)$ along $\bar{r}(4)^{-1}\pi^{-1}(\Delta_n)$.

Explicitly, since $\pi\bar{r}(4)$ is a block bundle projection, $\bar{r}(4): \bar{r}(4)^{-1}\pi^{-1}(\Delta_n) \to \pi^{-1}(\Delta_n)$ is an s-equivalence of (n+3)-ads as well as a normal map

$$(\pi^{-1}(\Delta_n)(j)) = \pi^{-1}(\Delta_n(j)) = \pi^{-1}(\partial_j \Delta_n)$$

of degree 1, $0 \leq j \leq n$, and

$$\pi^{-1}(\Delta_n)(n+1) = \pi^{-1}(\Delta_n) \cap \partial_+ V.$$

Let $K = \bar{r}(4)^{-1}\pi^{-1}(\Delta_n)$.

Let $(\bar{\bar{r}}, D)$ be a normal cobordism of $\bar{r}(4)|K$ modulo $\partial K - K(n+1)$ in general and in Case I modulo ∂K. We can then let $C' = C \cup_K D$, $r' = \bar{r} \cup \bar{\bar{r}}$ and adjust parameters to get a normal cobordism $r': C' \to (I \times E, I \times \mathcal{E})$ of $\bar{r}$. Further if $\bar{\bar{r}}(n+3): D(n+3) \to \pi^{-1}(\Delta_n)$ is an s-equivalence of (n+3)-ads, then $\pi r'(4)$ is a block bundle projection.

We now complete the proof of Theorem 5.1.0.

a) $\partial_- W = \emptyset$. Then $\bar{r}: C \to (I \times E)$ is an s-equivalence on $\partial C - C(2)$. There is a well-defined surgery obstruction to finding a cobordism of $\bar{r}$ modulo $\partial C - C(1)$ to an s-equivalence. This obstruction lies in $L_r(i)$, $i: \pi_1(E_+) \to \pi_1(E)$, where L denotes the Wall obstruction group. However, our assumptions e, f guarantee that $L(i) = L(j)$ when $j: \pi_1(\pi^{-1}(\Delta_n) \cap E_+) \to \pi_1(\pi^{-1}(\Delta_n))$, Δ_n a simplex of W of maximal dimension with $\Delta_n \cap \partial W = \emptyset$. Letting $K = \bar{r}(4)^{-1}\pi^{-1}(\Delta_n)$, we can find a normal cobordism $r': D \to I \times \pi^{-1}(\Delta_n)$ of $\bar{r}(4)|K$ whose surgery obstruction is the negative of the given one. Performing the interior pasting as in 1, we can let $C' = C \cup D$, $\bar{r}' = \bar{r} \cup r'$. Then we have cancelled the surgery obstruction. That is, we can find a normal cobordism of $\bar{r}'$ modulo $\partial C' - C'(2) = \partial C - C(2)$ to an s-equivalence $(\bar{\bar{r}}, \bar{C})$, $\bar{\bar{r}}: \bar{C} \to I \times E$. Note that if $\partial_+ V = \emptyset$ or if we are in Case I, the obstruction actually lies in $L(\pi_1(E))$. Then we can do the construction modulo ∂C_- so that this works in this case also. Our dimension assumptions allow us now to apply the s-cobordism theorem and deduce that $\overline{C} = I \times V$. Thus Theorem 5.1.0 is established if $\partial_- W = \emptyset$.

b) $\partial_- W \neq \emptyset$. Here the argument is more elementary. We first transform $(\overline{r}, C)$ to $(\overline{\overline{r}}, \overline{C})$ such that $\overline{\overline{r}}(2): \overline{C}(2) \to I \times E_+$ is an s-equivalence. We can do this through relative surgery as follows: (We can assume we are in Case II, otherwise there is nothing to do.) Since our hypotheses ensure that

$$\pi_1(I \times (E_+ \cap E_-)) \cong \pi_1(I \times E_+),$$

we can apply the fundamental theorem of surgery theory to first construct a relative normal cobordism of $\overline{r}(2): C(2) \to I \times E_+$ modulo $\partial C(2) - C(2)(1)$ to an s-equivalence of (n+5)-ads [18]. We can then extend this to a cobordism of $(\overline{r}, C)$ to $(\overline{\overline{r}}, \overline{C})$ such that $\overline{\overline{r}}(2): \overline{C}(2) \to I \times E_+$ is an s-equivalence. We now do relative surgery on $\overline{\overline{r}}$ modulo $\overline{C}(0) \cup \overline{C}(2) \cup \overline{C}(3) \cup \overline{C}(4)$ to get $\overline{\overline{\overline{r}}}, \overline{\overline{C}}$ such that $\overline{\overline{\overline{r}}}: \overline{\overline{C}} \to I \times E$ is an s-equivalence. We can do this since $\pi_1(I \times E_-) \cong \pi_1(I \times E)$. Now we can apply the s-cobordism theorem to yield the result. Q.E.D.

Section 2.

In this section we construct the fibrations necessary for Thoerem 5.0.1 and prove the first paragraph of that theorem.

For a manifold X with boundary we define the tangent bundle $\tau(X)$ of X to be the tangent bundle of the open extension of X, $X \cup \partial X \times [0,1)$, restricted to X. $\tau(\partial X)$ is a subbundle of $\tau(X)|\partial X$ and in fact $\tau(X)|\partial X = \tau(\partial X) \oplus \ell$, where ℓ is the trivial line bundle. The splitting is not canonical but any two are isotopic in the strongest possible sense. The splittings correspond to embeddings $\partial X \times [0,1] \hookrightarrow X$ extending $\partial X \hookrightarrow X$, i.e., collarings of ∂X. Any two are isotopic in the strongest possible sense [37] and this fact is known as the uniqueness of collarings. In the $p\ell$ case, absolute precision would demand we speak of micro-bundles and germ maps [26]. However, we ignore this complication since it presents only notational difficulties.

We now introduce some notation, conventions and a number of basic definitions.

Let X be an $(n+m)$-ad and Y an $(n+k)$-ad. By an n-ad map $f: X \to Y$, we mean a map of n-ads, where X, Y are considered as n-ads by taking the first $n-1$ faces.

Let E be a vector bundle (p or linear) over a manifold m-ad X. We give the vector bundle $\tau(X) \oplus E$ an m-ad structure

$$\tau(X) \oplus E(\alpha) = \tau X(\alpha) \oplus E[X(\alpha)],$$

where we use the notation $E[X(\alpha)]$ for E restricted to $X(\alpha)$. $\tau(X(\alpha)) \oplus E[X(\alpha)]$ is both a subspace of $\tau(X) \oplus E$ and a vector bundle over $X(\alpha)$.

Let X be an $(m+2)$-ad manifold, A an $(n+2)$-ad manifold, $f: X \to A$ an $(n+2)$-ad map. An n normal structure on f consists of

a. a vector bundle (p or linear depending on our category) R over X. R is called the bundle of the structure.

b. an (n+2)-ad bundle map ψ convering f

$$\begin{array}{ccc} \tau(X) \oplus R & \xrightarrow{\psi} & \tau(A) \oplus k\ell \\ \downarrow & & \downarrow \\ X & \xrightarrow{f} & A \end{array}$$

i.e., ψ is an isomorphism on each fiber, covers f and

$$\tau(X(\alpha)) \oplus R[X(\alpha)] \xrightarrow{\psi(\alpha)} \tau(A(\alpha)) \oplus k\ell\,[A(\alpha)],$$

where ℓ is the trivial line bundle over A. ψ is called the bundle map of the structure, A is called the indexing manifold of the structure. Often it will be an n-simplex. If we identify an n-normal structure with its stabilization gotten by replacing R by $R \oplus r\ell$, k by $k+r$, ψ by $\psi \oplus (f \times id)$, we get a stable n-normal structure.

Now let X be an (n+r+2)-ad manifold, Y an (n+r+2)-ad P. D. space of the same formal dimension as X, A an (n+2)-ad manifold, f: Y → A an (n+2)-ad map. Then a normal map of type (r, n) from X to Y over f, $(\lambda, \tilde{\lambda}): X \to (Y, \varepsilon)$, is

a) a vector bundle ε over Y

b) an (n+r+2)-ad map $\lambda : X \to Y$ of degree 1, in the usual sense [18]

c) a stable n-normal structure on $f\lambda$ with bundle $\lambda^*(\varepsilon)$ and bundle map

$$\tilde{\lambda} : \tau(X) \oplus \lambda^*(\varepsilon) \to \tau(A) \oplus k\ell\,.$$

Again A is called the indexing manifold, ε the bundle, $\tilde{\lambda}$ the bundle map of the normal map.

A normal map of type (r, n) is thus a fairly complicated object involving three spaces, three maps and three vector bundles. However, these turn out to be the key building blocks for constructing the fibrations which are at the heart of our work. We could simplify this notion slightly but at the cost of introducing a lot of cumbersome notation and machinery, a lot of unnaturally complicated proofs, yet the final result would be no more general. It appears these objects are the "natural" constituents for this theory.

Recall that we have an s-equivalence $r: (V, \partial_F V, \partial_+ V, \partial_- V) \to (E, E_F, E_+, E_-)$, where $\pi: (E, E_+) \to W$ is a Hurewicz fibration, $E_- = \tau^{-1}(\partial_- W), E_F = \tau^{-1}(\partial_F W)$, and $f = \tau r$. For our constructions in this section we are going to dispense with our assumptions on f and consider a more general situation.

Let $\pi: (E, E_+) \to W$ be an arbitrary pair of Serre fibrations over a connected, simply connected m-ad manifold. π is called a <u>PD fibration</u> of fiber dimension d if for any $x \in W$ and hence for each $x \in W$ we are given an s-structure on $(\pi^{-1}(x), \pi_+^{-1}(x))$ which make it into a finite PD 2-ad of formal dimension d. Note that E has a natural (m+1)-ad structure with $E(i) = \pi^{-1}(W(i))$, $0 \leq i \leq m-2$, $E(m-1) = E_+$. We say we are in Case I if $\pi_+: E_+ \to W$ is a fiber bundle with fiber a closed manifold whose s-structure is the given one on $(\pi^{-1}(x), \pi_+^{-1}(x))$. Otherwise we are in Case II.

<u>Remark.</u> If we do not assume W is simply connected, a finite PD s-structure on $(\pi^{-1}(x), \pi_+^{-1}(x))$ does not necessarily impose a unique such structure on $(\pi^{-1}(y), \pi_+^{-1}(y))$. It is to avoid this complication that we assume W is both connected and simply connected. Notice that for a PD fibration of fiber dimension d, $\pi^{-1}(\Delta_n)$ is a PD (n+3)-ad of formal dimension d+n.

<u>Remark.</u> We could easily generalize to the situation where π is a k-ad fibration $\pi: (E, E_0, \ldots, E_{k-1}) \to W$ rather than a 2-ad fibration.

Now let ε be a bundle over E. We will construct the following Δ fibrations [28] over W: $A_1(\pi, \varepsilon) \subset A_2(\pi, \varepsilon) = A_2(\pi)$.

An n-simplex of $A_2(\pi)$ is

a) an n-simplex Δ_n of W,

b) a normal map $(\lambda, \tilde{\lambda}): X \to (\pi^{-1}(\Delta_n), \varepsilon)$ of type $(1, n)$ of X to $\pi^{-1}(\Delta_n)$ over $\pi | \pi^{-1}(\Delta_n)$. If we are in Case I we assume $\lambda(n+1): X(n+1) \to \pi_+^{-1}(\Delta_n)$ is an isomorphism.

The face operators are $\partial_i(\lambda, \tilde{\lambda}) = (\lambda(i), \tilde{\lambda}(i)): X(i) \to (\pi^{-1}(\partial_i(\Delta_n), \varepsilon))$, $0 \leq i \leq n$. $A_1(\pi) \subset A_2(\pi)$ is the sub Δ-set of those simplices such that λ is an s-equivalence of $(n+3)$-ads.

There is a natural Δ map $\rho_2: A_2(\pi) \to W$. Let $\rho_1 = \rho_2 | A_1(\pi)$. We wish to establish basic properties of these fibrations. They are all consequences of the basic amalgamation property of maps of degree $(1, n)$. It is important to understand this property. We begin with the simplest and model example.

Suppose X and Y are manifolds of dimension d, f is an equivalence between $A \subset \partial X$ and $B \subset \partial Y$, where A, B are submanifolds of dimension d-1 of $\partial X, \partial Y$, respectively. Then one can glue X to Y by f, getting $X \cup_f Y$, also a manifold of dimension d. A complication arises in the smooth category since the smooth structure on $X \cup_f Y$ is not uniquely determined by the given data but depend also on choosing collars of ∂A in ∂X, ∂X in X, ∂B in ∂Y and ∂Y in Y. However, the smooth structure on $X \cup_f Y$ is determined up to diffeomorphism isotopic to the identity [31]. This gluing is the model for amalgamations. If X is an n-ad manifold, $A = X(i)$, Y an m-ad manifold, $B = Y(j)$, then $X \cup_f Y$ is an $(n+m-3)$-ad manifold having as faces the faces of X other than A and the faces of Y other than B.

The above construction goes through with less technical annoyance when X and Y are n-, m-ad spaces rather than manifolds.

There is another sort of amalgamation appropriate to n-ads. If X is an (n+2)-ad then we can form an (n+3)-ad by amalgamating two of the codimension one faces. For convenience this amalgamated subspace of codimension one is thought of as the last codimension one face of our (n+1)-ad. Of course, we can iterate this to amalgamate any subset of the family of codimension one faces. This operation restricts to an operation in the category of manifold ads,provided in the smooth case, we permit smooth manifolds with corners along the boundary and require that faces meet transversally at intersections [A1]. In the situation of manifold n-ads, amalgamation of codimension one faces is just the amalgamation along their intersection as described in previous paragraph.

Using the type of amalgamation discussed above we can form an amalgamation in the following situation.

Suppose we are given $X_0, \dots, X_p$, where each X_i is an $(2+n_i+r)$-ad and suppose for some indices i, j, k, where $k_j \leq n_i$, $i_j \leq n_k$, we are given $g^j_{ik}: X_i(k_j) \cong X_k(i_j)$, where we assume $i < k$. We further assume that each codimension one face of X_t show up at most once as some $X_t(s_j)$. Write $\bar{g} = \{g^j_{i,k}\}$, which we call an amalgamating family of functions. We can amalgamate all the X_i via elements of $\bar{g}$ to form $\bigcup_{\bar{g}} X_i$. If each X_i is a d-dimensional $(2+n_i+r)$-ad manifold or P.D. space, then $\bigcup_{\bar{g}} X_i$ is a d-dimensional t-ad manifold or PD space, the t depending on the number of identificactions $\{g^j_{i,k}\}$. In general, $\bigcup_{\bar{g}} X_i$ is a (p+2)-ad space with $(\bigcup_{\bar{g}} X_i)(j) = X_j$. We are more concerned with the (p+r+3)-ad spaces $I \times \bigcup_{\bar{g}} X_i$:

$$I \times \bigcup_{\bar{g}} X_i(j) = 0 \times X_j \qquad 0 \leq j \leq p$$

$$I \times \bigcup_{\bar{g}} X_i(p+t) = I \times \bigcup_i X_i(n_i+t) \qquad 1 \leq t \leq r$$

$$I \times \bigcup_{\bar{g}} X_i(p+r+1) = (1) \times \bigcup_{\bar{g}} X_i \cup I \times \bigcup_{s \neq t} X_0(t) \;,$$

where $X_s(t)$ are codimension one faces of X_s, $0 \leq t \leq n_s$ which are not in the range or domain of any of the $g^j_{i,k}$. If the X_i are all (n_i+r+2)-ad manifolds or PD spaces of dimension d, then $I \times \bigcup_g X_i$ is a $(p+r+3)$-ad manifold or PD space of dimension d+1.

The following picture illustrates the construction with $p = 1$, $X_0 = X_1 = I$, $r = 0$.

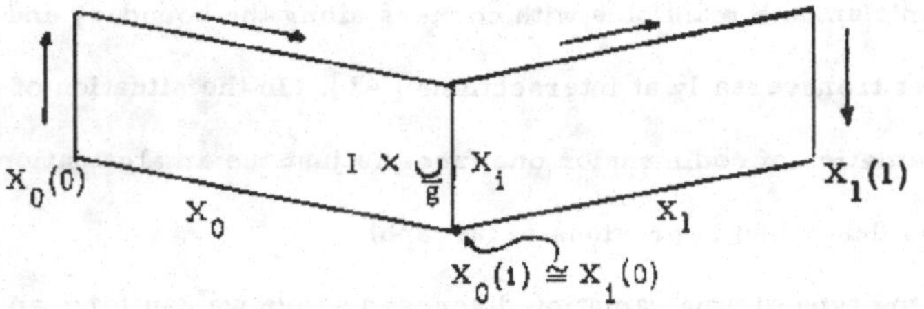

The arc running from $X_0(0)$ to $X_1(1)$ along $\longrightarrow$ is $I \times \bigcup_g X_i(2)$.

The crucial fact is that this type of amalgamation can be extended to maps of type (r, n). More explicitly, we have the following proposition whose statement is more difficult than its proof, which is routine.

<u>Proposition 5.2.0.</u> Suppose we have normal maps of type (r, n_i) from X_i to Y_i over $f_i : (\lambda_i, \tilde{\lambda}_i)(X_i) \to (Y_i, \varepsilon_i)$ with $f_i : X_i \to A_i$. Suppose we are given amalgamating families of functions

$$g^j_{i,k} : A_i(k_j) \approx A_k(i_j)$$
$$p^j_{i,k} : X_i(k_j) \approx X_k(i_j)$$
$$\Omega^j_{i,k} : Y_i(k_j) \approx Y_k(i_j)$$
$$\rho^j_{i,k} : \varepsilon_i[Y_i(k_j)] \approx \varepsilon_k[Y_k(i_j)]$$

where $\rho^j_{i,k}$ are also bundle equivalences. Consider the differentials

$$dg^j_{i,k}: t(A_i(K_j)) \approx \overline{t}(A_k(i_j))$$

$$dp^j_{i,k}: t(X_i(k_j)) \approx t(X_k(j_j)) .$$

We require all reasonable commutativity relation to hold, i.e.,

$$\lambda_k(i_j)p^j_{i,k} = Q^j_{i,k}\lambda_i(k_j)$$

$$f_k(i_j)Q^j_{i,k} = g^j_{i,k}f_i(k_j)$$

$$\pi_k\rho^j_{i,k} = Q^j_{i,k}\pi_i \quad \text{where } \pi_s: \mathcal{E}_s \to Y_s \text{ is the bundle projection.}$$

The maps $\rho^j_{i,k}$, $Q^j_{i,k}$, $p^j_{i,k}$ and the commutativity relations define

$$\psi^j_{i,k}: \lambda^*_i(\varepsilon_i)[X_i(k_j)] \simeq \lambda^*_k[\varepsilon_k[X_k(i_j)]]$$

we require

$$\widetilde{\lambda}_k(i_j)(dp^j_{i,k} \oplus \psi^j_{i,k}) = (dg^j_{i,k} \oplus \mathrm{id})\widetilde{\lambda}_i(k_j).$$

Then:

I. We can amalgamate $(\lambda_i, \widetilde{\lambda}_i): X_i \to (Y_i, \varepsilon_i)$ to $(\lambda', \widetilde{\lambda}'): \bigcup_{\overline{P}} X_i \to (\bigcup_{\overline{Q}} Y_i, \bigcup_{\overline{P}} \mathcal{E}_i)$, which is a normal map over the amalgamation of f_i:

$$f': \bigcup_{\overline{Q}} Y_i \to \bigcup_{\overline{g}} A_i$$

II. We have the natural extensions of $(\lambda', \widetilde{\lambda}')$ to $(\lambda, \widetilde{\lambda}): I \times \bigcup_{\overline{P}} X_i \to (I \times \bigcup_{\overline{Q}} Y_i, \mathcal{E})$, where $\mathcal{E} = I \times \bigcup_{\overline{P}} \varepsilon_i$, a normal map of type $(r, p+1)$ over $f: I \times \bigcup_{\overline{Q}} Y_i \to I \times \bigcup_{\overline{g}} A_i$, where $f = \mathrm{id} \times f'$.

The application we have in mind of the above is when each A_i is a p-simplex and $\bigcup_{\overline{g}} A_i \cong \partial B - \partial_j B$, where B is a $(p+1)$-simplex. Then we can identify $I \times \bigcup_{\overline{g}} A_i \simeq B$ as a $(p+3)$-ad.

Let K be any simplicial complex, $s \geq 0$. Define the following Δ set $\bigcup(r, s, K)$ as follows.

An n-simplex of $\cup(r,s,K)$ is

a) an n-simplex Δ_n of K,

b) a normal map $(\lambda,\tilde{\lambda}): X \to (Y,\varepsilon)$ of type (r,n) from X to Y over f, where $f: Y \to \Delta_n$ and X and Y have formal dimension $n+s$. The face operations are defined by $\partial_i(\lambda,\tilde{\lambda}) = \lambda(i),\tilde{\lambda}(i): X(i) \to (Y(i), \varepsilon(Y(i)))$, $0 \leq i \leq n$, which is a normal map of degree (r,n) over $f(i): Y(i) \to \partial_i\Delta_n$.

There is a natural Δ map $\rho: \cup(r,s,K) \to K$.

Remark. The Δ fibrations over W we will construct are all subfibrations of $\rho = \rho(r,s)$. In particular, $A_1(\pi) \subset A_2(\pi) \subset \cup(1,d,W)$.

Proposition 5.2.1. ρ is a Δ Kan fibration [ ].

Proof. This is an immediate consequence of 5.2.0 since if $A_i = \partial_i B$, B a $(p+1)$-simplex of K, then the construction of 5.2.0 and the identification of $I \times \bigcup_g A_i = B$ yields a $(p+1)$-simplex of $\cup(r,s,K)$ whose faces are the given p-simplices of $\cup(s,K)$ over the A_i.

Theorem 5.2.2. $\rho_1: A_1(\pi) \to W$, $\rho_2: A_2(\pi) \to W$ are Kan fibrations.

Proof. Observe that $A_1(\pi) \subset A_2(\pi) \subset \cup(1,d,W)$, $\rho | A_2(\pi) = \rho_2$. Let Δ_{p+1} be a $(p+1)$-simplex of W and $\Delta_p^i = \partial_i\Delta_{p+1}$, $0 \leq i \leq p$. Then the $g_{i,k}^j$ are defined by the common faces of Δ_p^i and Δ_p^j.

Let $(\lambda_i,\tilde{\lambda}_i): X_i \to (Y_i, \varepsilon_i) \in \rho_2^{-1}(\Delta_p^c)$, $0 \leq c \leq p$, where $\varepsilon_i = [Y_i]$, which agree over common faces. Then this provides the $P_{i,k}^j, Q_{i,k}^j, \rho_{i,k}^j$ and the coherence relations of 5.20. We can then form

$$(\lambda,\tilde{\lambda}): I \times \bigcup_{\overline{P}} X_i \to (I \times \bigcup_{\overline{Q}} Y_i, \varepsilon) \text{ over } f: I \times \bigcup_{\overline{Q}} Y_i \to I \times \bigcup_{\overline{g}} \Delta_p^i .$$

However, since $\pi^{-1}(\Delta_p) \to \Delta_p$ is a Serre fibration, we have $(p+1)$-ad maps

$$\begin{array}{ccc} I \times \bigcup_{\overline{Q}} Y_i & \xrightarrow{\eta} & \pi^{-1}(\Delta_{p+1}) \\ \downarrow & & \downarrow \\ I \times \bigcup_{\overline{g}} \Delta^i_p & \xrightarrow{\gamma} & \Delta_{p+1} \end{array}$$

which are the identity on Y_i and Δ^i_p such that γ is an isomorphism and we also have a bundle map $\tilde{\eta}: I \times \bigcup_{\rho} \varepsilon_i \to \varepsilon[\pi^{-1}(\Delta_{p+1})]$ over η. Here

$$\lambda^*(I \times \bigcup_{\rho} \varepsilon_i) = (\eta\lambda)^*(\varepsilon[\pi^{-1}(\Delta_{p+1})]) = (\eta\lambda^*)(\varepsilon).$$

Then $(\eta\lambda, d\gamma\lambda): I \times \bigcup_{\overline{Q}} X_i \to (\pi^{-1}(\Delta_{p+1}), [\pi^{-1}(\Delta_{p+1})])$ is a (p+1)-simplex of $A_2(\pi)$ with $\partial_i(\eta\lambda, d\gamma\tilde{\lambda}) \equiv (\lambda_i, \tilde{\lambda}_i)$. If $(\lambda_i, \tilde{\lambda}_i) \in A_1(\pi)$ then so is $(\eta\lambda, d\gamma\tilde{\lambda})$. Q.E.D.

Now let $\psi: W \to A_2(\pi)$ be an cross section of ρ_2. Then for each simplex $\Delta^\alpha_r \in W$ of maximal dimension,

$$\psi(\Delta^\alpha_r) = (\lambda^\alpha, \tilde{\lambda}^\alpha): X^\alpha \to (\pi^{-1}(\Delta^\alpha_r), \varepsilon[\pi^{-1}(\Delta^\alpha_r)]$$

is a map of type (1, r) over π. Again the common faces provide the amalgamation functions $g^j_{\alpha\lambda}: \Delta^\alpha_r(\gamma_j) \cong \Delta^\alpha_r(\alpha_j)$ while the fact that ψ is a cross section supplies the $p^j_{\alpha\gamma}$, $Q^j_{\alpha\gamma}$, $\rho^j_{\alpha\gamma}$ and the coherence relations.

Thus by 5.2.0, we have a normal map of type (1, m)

$$(\hat{\psi}, \hat{\hat{\psi}}): \bigcup_{\overline{P}} X^\alpha \to (\bigcup_{\overline{Q}} \pi^{-1}(\Delta^\alpha_r), \bigcup_{\overline{\rho}} \varepsilon|)$$

over

$$\pi^{-1}: \bigcup_{\overline{Q}} \pi^{-1}(\Delta^\alpha_r) \to \bigcup_{\overline{g}} \Delta^\alpha_r .$$

However, $\bigcup_{\overline{Q}} \pi^{-1}(\Delta^\alpha_r) = E$, $\bigcup_{\overline{g}} \Delta^\alpha_r = W$, $\pi| = \pi$, $\bigcup_{\overline{\rho}} \varepsilon| = \varepsilon$. Thus we have a normal map of type (1, m-2)

$$(\hat{\psi}, \hat{\hat{\psi}}): \bigcup_{\overline{P}} X^\alpha \to (E, \varepsilon)$$

over $\pi: E \to W$. Here $\hat{\psi}: \bigcup_{\overline{P}} X^\alpha \to E$ is an (m+1)-ad map of degree 1.

Clearly, $\pi\hat{\psi}$ is a block bundle projection if and only if $\psi(W) \subset A_1(\pi)$. Further, if we are in Case I, $\hat{\psi}: \bigcup_{\overline{p}} X^{\alpha}(m-1) \simeq E(m-1) = E_r$. Thus we have proved

Proposition 5.2.3. To each cross section ψ of $A_2(\pi) = A_2(\pi, \varepsilon)$ there corresponds a normal map of type $(1, m-2)$

$$(\hat{\psi}, \hat{\hat{\psi}}): \bigcup_{\overline{p}} X^{\alpha} \to (E, \varepsilon) \quad \text{over} \quad \pi: E \to W.$$

Here $\hat{\psi}$ is an $(m+1)$-ad map of degree 1. If we are in Case I,

$$\hat{\psi}(m-1): \bigcup_{\overline{p}} X^{\alpha}(m-1) \to E_+$$

is an isomorphism. Further, $\pi\hat{\psi}$ is a block bundle projection if and only if $\psi(W) \subset A_1(\pi)$. $(\hat{\psi}, \hat{\hat{\psi}})$ is called the amalgamation of ψ.

We now wish to link-up homotopy classes of cross sections and cobordism classes of normal maps. The correspondence is quite neat. Before stating it, we need some definitions.

We consider $\mathrm{id} \times \pi: I \times E \to I \times W$. Then if $I \times W$ is given a simplicial subdivision which agrees with the given structure on W on the two ends, we can identify $A_i(\mathrm{id} \times \pi, I \times \varepsilon) | (t) \times W = A_i(\pi, \varepsilon)$, $i = 1, 2$, $t = 0, 1$. For any fibration $g: X \to Y$ we denote the crossection by $\Gamma(g)$ or $\Gamma(X)$.

We say $\psi_1, \psi_2 \in \Gamma(A_2(\pi, \varepsilon))$ are homotopic, $\psi_1 \sim \psi_2$, if and only if for some subdivision of $I \times W$ agreeing with the given one on both ends, there exists a $\psi \in \Gamma(A_2(\mathrm{id} \times \pi, I \times \varepsilon)$ such that $\psi | 0 \times W = \psi_1$ and $\psi | 1 \times W = \psi_2$. If

$$(\lambda_1, \tilde{\lambda}_1): X_1 \to (Y, \varepsilon) \quad \text{and} \quad (\lambda_2, \tilde{\lambda}_2): X_2 \to (Y, \varepsilon)$$

are normal maps of type (r, n) over $f: Y \to A$, then a normal cobordism of $(\lambda_1, \tilde{\lambda}_1)$ to $(\lambda_2, \tilde{\lambda}_2)$ is a normal map of type $(r, n+2)$, $(\lambda, \tilde{\lambda}): C \to (I \times Y, I \times \varepsilon)$ over $\mathrm{id} \times f: I \times Y \to I \times A$, where C is a cobordism of X_1 to X_2 and the restriction of $(\lambda, \tilde{\lambda})$ to X_i is $(\lambda_i, \tilde{\lambda}_i)$, $i = 1, 2$. $(\lambda, \tilde{\lambda})$ is a normal cobordism modulo i if C is a cobordism modulo (i) and $(\lambda(i), \tilde{\lambda}(i)) = (\mathrm{id} \times \lambda_1(i), \tilde{\lambda}_1(i))$, $0 \leq i \leq n+r+2$.

Proposition 5.2.4. Every normal map $(\lambda, \tilde{\lambda}): X \to (E, \varepsilon)$ over π (where $\lambda(n+1)$ is an isomorphism in Case I) is homotopic to one of the form $(\hat{\psi}, \hat{\hat{\psi}})$, $\psi \in \Gamma A_2(\pi)$. Further, the correspondence from crossections of $A_2(\pi)$ to their amalgamations induces a bijection from homotopy classes of crossections to normal cobordism classes (modulo $(m-1)$ in Case I) of maps of type $(1, m-2)$ over π (with $\lambda(m-1)$ an isomorphism in Case I).

Proof. Suppose we are given $\psi \in \Gamma(A_2(\mathrm{id} \times \pi, I \times \varepsilon))$, then $(\hat{\psi}, \hat{\hat{\psi}})$ is a normal cobordism between $(\hat{\psi}_0, \hat{\hat{\psi}}_0)$ and $(\hat{\psi}_1, \hat{\hat{\psi}}_1)$, where $\psi_t = \psi | (t) \times W$, $t = 0, 1$. Thus homotopic crossections induce normally cobordant maps. We will prove both statements of the proposition by constructing an inverse to the correspondence induced by amalgamation.

We have the commutative diagram

$$\begin{array}{ccc} \tau(X) \oplus \lambda^*(\varepsilon) & \xrightarrow{\tilde{\lambda}} & \tau(W) \oplus k e \\ \downarrow & & \downarrow \\ X & \xrightarrow{\pi\lambda} & W \end{array}$$

with $\tilde{p}\tilde{\underline{0}} = $ identity.

We also have bundle maps

$$\begin{array}{ccccc} \tau(X) \oplus \lambda^*(\varepsilon) & \xrightarrow{\tilde{\underline{0}}} & \tau(\lambda^*(\varepsilon)) & \xrightarrow{\tilde{p}} & \tau(X) \oplus \lambda^*(\varepsilon) \\ \downarrow & & \downarrow & & \downarrow \\ X & \xrightarrow{\underline{0}} & \lambda^*(\varepsilon) & \xrightarrow{p} & X \end{array}$$

where $\underline{0}$ is the 0-section. Since $\tau(W) \oplus ke = i^*(\tau(W) \times R^k))$, where $i: W \to W \times R^k$ is the 0-section, composing the various maps we have

$$\begin{array}{ccccc} \tau(\lambda^*(\varepsilon)) & \xrightarrow{\tilde{i}\tilde{\lambda}\tilde{p}} & \tau(W \times R^k) & \xrightarrow{\tilde{\pi}_1} & \tau(W) \oplus ke \\ \downarrow & & \downarrow & & \downarrow \\ \lambda^*(\varepsilon) & \xrightarrow{i\pi\lambda p} & W \times R^k & \xrightarrow{p\pi_1} & W \end{array}$$

By the fundamental theorem of immersion theory [ ], $i\pi\lambda p \to i\tau\lambda p$ can be deformed to an immersion γ and this deformation can be covered by a deformation of $i\lambda p$ to $d\gamma$, the differential of γ. Now if k is large with respect to the dimension of X, γ can be taken to be an embedding on X, thus on a neighborhood of X (since it is an immersion) and finally on all of $\lambda^*(\varepsilon)$ since a bundle can be shrunk into any neighborhood of its 0-section. Further we can assume that γ restricted to $\overline{0}(X)$ is transverse to $\Delta^\alpha \times R^k$ for each simplex Δ^α of W. Now we observe that the deformation of $\pi_1(\pi\lambda p\underline{0}) = \pi\lambda$ to $\pi_1\gamma\underline{0}$ can be covered by a deformation of λ to λ_1 because of the covering homotopy property of $\pi: E \to W$. Thus we have $\lambda_1 \sim \lambda$, $\lambda_1: X \to E$.

Let $X^\alpha = (\pi\lambda_1^{-1})(\Delta^\alpha) \cong \gamma\underline{0}(X) \cap \Delta^\alpha \times R^k$. Since $\gamma\underline{0}$ is transversal to $\Delta^\alpha \times R^k$, $\lambda_1^\alpha = \lambda_1 | X^\alpha : X^\alpha \to \pi^{-1}(\Delta^\alpha)$ is an $(|\alpha| + 3)$-ad map of degree 1. (If we are in Case I, we must and can choose γ, λ_1 carefully so that $\lambda_1^\alpha | X^\alpha \cap X(M-1) = \lambda | X^\alpha \cap X(m-1)$.)

We now have the following diagram.

$$\begin{array}{ccccccc}
\tau(X) \oplus \lambda^*(\varepsilon) & \xrightarrow{\underline{0}} & \tau(\lambda^*(\varepsilon)) & \xrightarrow{d\gamma} & \tau(W \times R^k) & \xrightarrow{\pi_1} & \tau(W) \oplus ke \\
\downarrow & & \downarrow & & \downarrow & & \downarrow \\
X & \xrightarrow{\underline{0}} & \lambda^*(\varepsilon) & \xrightarrow{\gamma} & W \times R^k & \xrightarrow{\pi_1} & W
\end{array}$$

where $\lambda^*(\varepsilon) \approx \lambda_1^*(\varepsilon)$ since λ_1 and λ are homotopic

Since $\gamma\underline{0}$ is transversal to each $\Delta^\alpha \times R^k$ and since $\lambda^*(\varepsilon)$ is a normal bundle for $\gamma\underline{0}(X)$ in $W \times R^k$, we can deform $d\gamma\underline{0}$ to W through bundle maps over $\gamma\underline{0}$ such that for each X^α

$$\begin{array}{ccccc}
\tau(X^\alpha) \oplus \lambda^*(\varepsilon)[X^\alpha] & \longrightarrow & \tau(\Delta^\alpha \times R^k) & \hookrightarrow & \tau(W \times R^k) \\
\downarrow & & \downarrow & & \downarrow \\
X^2 & \xrightarrow{\gamma\underline{0}} & \Delta^\alpha \times R^k & \subset & W \times R^k
\end{array}$$

Let $\tilde{\lambda}_1 = \tilde{\pi}_1\omega$ and $\tilde{\lambda}_1^{\alpha} = \tilde{\lambda}_1 \mid \tau(\lambda^2) \oplus \lambda^{\alpha}(\varepsilon)[X^{\alpha}] = \pi_1$. Then we have $(\lambda_1^{\alpha}, \tilde{\lambda}_1^{\alpha}): (X^{\alpha}) \to (\pi^{-1}(\Delta^{\alpha}), \varepsilon | \pi^{-1}(\Delta^{\alpha}))$ is a normal map of type $(1, m)$ over π. This defines a crossection of ρ_2, i.e., an element ψ of $\Gamma(A_2, \pi)$. Clearly, $(\hat{\psi}, \hat{\hat{\psi}}) = (\lambda_1, \tilde{\lambda}_1)$. Hence we have established the first sentence of the proposition. Let us call the construction from normal maps of type $(1, m)$ over π to crossections of $A_2(\pi)$ a <u>dissection</u> of the normal map. If ψ, ψ are crossections such that $(\hat{\psi}, \hat{\hat{\psi}})$ and $(\hat{\psi}', \hat{\hat{\psi}}')$ are normally cobordant, then dissecting the cobordism so as to extend the given dissection on the two ends yields a homotopy of ψ to ψ'. Since amalgamation and dissection are inverse operations, the proof is complete.

<u>Corollary 5.2.5</u>. A normal map $(\lambda, \tilde{\lambda}): X \to (E, \varepsilon)$ over π is cobordant to $(\lambda', \tilde{\lambda}')$, with $\pi\lambda'$ the projection of a block bundle, if and only if some and hence every dissection of $(\lambda, \tilde{\lambda})$, $\psi \in A_2(\pi)$, is homotopic to $\psi' \in A_1(\pi)$.

It is possible and convenient to replace the problem of deforming a crossection of a fiber space to one of a sub fiber space by the problem of finding a crossection to an associated fiber space. This is a convenient device which allows one to analyze the obstruction directly.

The general construction is as follows. Suppose we are given a commutative diagram of spaces or (Δ sets)

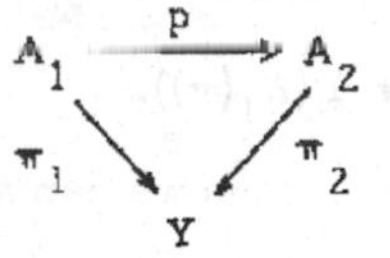

where π_1, π_2 are Serre (Kan) fibrations. First replace p by the associated Serre (Kan) fiber map $\overline{p}: \overline{A}_1 \to A_2$. Given a crossection ψ of π_2, we have a pullback fibration $\overline{\overline{p}}: \psi^*(\overline{A}_1) \to Y$. Then ψ is homotopic to a crossection of π_1 if and only if $\overline{\overline{p}}$ has a crossection. In our situation we will give a direct construction of $\psi^*(\overline{A}_1)$, $\overline{\overline{p}}$ since we wish to examine the homotopy group of $\overline{\overline{p}}$ homotopy group of $\overline{p}$.

Suppose we are given a crossection ψ of $A_2(\pi)$. We construct a Δ set $A_\psi(\pi)$ as follows

An n-simplex of $A_\psi(\pi)$ is an n-simplex Δ_n of W and a normal cobordism $(C,\tilde{C}): Z \to (I \times \pi^{-1}(\Delta_n), I \times \varepsilon)$ of $\psi(\Delta_n) = (\lambda, \tilde{\lambda}): X \to (\pi^{-1}(\Delta_n), \varepsilon|)$ such that $C(n+3): Z(n+3) \to (\pi^{-1}(\Delta_n))$ is a simple homotopy equivalence of (n+3)-ads. In Case I, we assume $(C,\tilde{C})$ is a normal cobordism modulo (n+1).

We define $\partial_i(C,\tilde{C}) = (C(i), C(i))$, turning $A_\psi(\pi)$ into a Δ set. There is a natural projection $\rho_\psi: A_\psi(\pi) \to W$ and further there is a Δ map, $\mu: A_\psi(\pi) \to A_1(\pi)$, where $\mu(C,\tilde{C}) = (C(n+3), \tilde{C}(n+3))$ and we have the commutative diagram

$$\begin{array}{ccc} A_\psi(\pi) & \xrightarrow{\ \mu\ } & A_1(\pi) \\ & {}_{\rho_\psi}\searrow \quad \swarrow_{\rho_1} & \\ & W & \end{array}$$

<u>Proposition 5.2.6.</u> ρ_ψ is a Kan fibration. (Warning: It is possible that $A_\psi(\pi) = \emptyset$.)

The proof proceeds as in 5.2.2.

Let $W' \subset W$ be a subcomplex and suppose ψ restricted to W' $\tau' \in \Gamma(A_1(\pi)|W')$. This induces a crossection $\lambda_{W'}$ of $A_\psi(\pi)|W'$, where $\lambda_{W'}(\Delta_n) = \mathrm{id} \times \psi(\Delta_n)$.

<u>Proposition 5.2.7.</u> $\lambda_{W'}$ extends to a crossection of $A_\psi(\pi)$ if and only if ψ is homotopic modulo W' to $\psi' \in \Gamma(A_1(\pi))$.

<u>Proof.</u> If $\lambda_{W'}$ extends to λ then we can amalgamate to yield $(\hat{\lambda}, \hat{\tilde{\lambda}}): C \to (I \times E, I \times \varepsilon)$, a normal map of type $(1, m)$ over $\mathrm{id} \times \pi: I \times E \to I \times W'$. This is a normal cobordism of $(\hat{\psi}, \hat{\tilde{\psi}})$ to $(\hat{\lambda}(m), \hat{\tilde{\lambda}}(m))$ with $\pi\hat{\lambda}(m)$ a block bundle projection. By a straightforward generalization of 5.2.5 to a relative version, we get ψ is homotopic modulo W' to $\psi' \in \Gamma(A_1(\pi))$.

Conversely, if h is a crossection of $A_2(\mathrm{id} \times \pi)$ over $I \times W$, where $I \times W$ is given some subdivision of the standard triangulation which agrees with the given one on the two ends, and if we further subdivide $I \times W$, modulo the two ends, we can use dissection to subdivide the crossection h to a crossection h', modulo the two ends, of the subdivided complex. Thus if $\psi \sim \psi'$ one can assume the homotopy h is defined on a subdivision of $I \times W$ such that $I \times \Delta^{\alpha}$ is a subcomplex of W. Thus given $h \in \Gamma A_2(\mathrm{id} \times \pi)$ and $\Delta^{\alpha} \in W$, define

$$(\bar{h}^{\alpha}, \bar{\bar{h}}^{\alpha}): X^{\alpha} \to (I \times \pi^{-1}(\Delta^{\alpha}), I \times \mathcal{E})$$

by amalgamating h over all simplices of $I \times W$ lying in $I \times \Delta^{\alpha}$. This defines the crossection λ of $A_{\psi}(\pi)$.

We can now pull together our results and complete the proof of the first paragraph of 5.0.1.

__Proposition 5.2.8.__ Suppose W is a 3-ad and $(r, \tilde{r}): V \to (E, \mathcal{E})$ is a map of degree $(1, 1)$ over $\pi: E \to W$. We write

$$W(0) = \partial_F W \qquad W(1) = \partial_- W$$
$$V(0) = \partial_F V \qquad V(1) = \partial_- V$$
$$V(2) = \partial_+ V$$

Suppose r is an s-equivalence and πr satisfies assumptions a)-h). Let ψ be a dissection of (r, r). Then there exists a crossection $\lambda_{\partial_F W}$ of $A_{\psi}(\pi)$ over $\partial_F W$ and r is homotopic to a block bundle projection if and only if $\lambda_{\partial_F (W)}$ extends to λ a crossection of $A_{\psi}(\pi)$ over W.

__Proof.__ $\lambda_{\partial_F W}$ as defined above extends to a crossection λ if and only if ψ is homotopic rel $\partial_F W$ to $\psi' \in \Gamma(A_1(\pi))$. ψ is homotopic to ψ' if and only if $(r, \tilde{r})$ is cobordant modulo $\partial_F V$ (and in Case I also $\partial_+ V$) to a block bundle projection by 5.2.5. By 5.1.0, (r, r) is cobordant to a block bundle projection if and only if r is homotopic (modulo $\partial_F V$ and in Case I also modulo $\partial_+ V$) to a block bundle projection. Q.E.D.

We have thus proved the first paragraph of 5.0.1, taking $A(f) = A_\psi(\pi)$, ψ a dissection of πf.

Remark. The notion of cobordism in Sections 1 and 2 are slightly different, (i.e. note that two different bundles over E are being used in the different sections). However, it is simple to prove (and we will leave it to the reader to do so) that cobordism in one sense implies cobordism in the other.

Let us now examine the situation when we do not make the troublesome assumption h) and see what we can conclude.

To get a result in this situation we weaken the notion of block bundle projection to quasi block bundle projection. We say that $f: V \to W$ is a quasi block bundle projection if f is a transverse map such that for each pair of simplices $\Delta_i \subset \Delta_j \subset W$, $(f^{-1}(\Delta_i), f_+^{-1}(\Delta_i)) \hookrightarrow (f^{-1}(\Delta_j), f_+^{-1}(\Delta_j))$ is a homotopy equivalence which is a simple homotopy equivalence when $\Delta_j \neq \Delta_m$, where Δ_m is a fixed simplex of W of maximal dimension. We also require that for some and hence every vertex p of W, $(f^{-1}(p), f_+^{-1}(p)) \to (L, \partial L)$ an s-equivalence.

Then one can prove

Theorem (5.1.0)'. Under assumptions a)-g), if r is cobordant to a block bundle projection, then r is homotopic to a quasi block bundle projection.

Using this we copy the proof of 5.2.8 to get

Proposition (5.2.8)'. Suppose W is a 3-ad and $(r, \tilde{r}): V \to (E, \mathcal{E})$ a map of type (1, 1) over $\pi: E \to W$. Suppose r is a homotopy equivalence and πr satisfies assumptions a)-g). Let ψ be a dissection of πr. If $\lambda_{\partial_F W}$ extends to λ a crossection of $A_\psi(\pi)$ over W, then f is homotopic to a quasi block bundle projection.

Since the second half of 5.0.1 does not involve assumption h), Corollaries 5.0.2, 5.0.3 and 5.0.4 hold in the absence of h) provided the phrase "block bundle projection" is replaced by "quasi block bundle projection".

Section 3.

In this section we complete the proof of Theorem 5.0.1. To do this, we must identify the obstruction to extending a crossection of $A(f) = A_{\psi}(\pi)$. As in the previous section, we consider the more general situation of a P.D. fibration $\pi: (E, E_+) \to W$, where W is a connected, simply connected m-ad manifold.

The difficulty in applying the usual obstruction theory is that the Δ set W does not satisfy a local Kan condition; i.e., each simplex does not lie in a sub-complex of W which is Kan. Since this condition seems necessary to ape the usual obstruction theory, at least in the obvious simple-minded way, we enlarge W as follows.

Let K be an ordered simplicial complex (vertices are ordered). By $\underset{\sim}{K}$ we mean the Δ set whose n simplices are simplicial order preserving maps $\tau: \Delta_n \to K$ where Δ_n is the standard n-simplex. Those τ corresponding to injections are just the simplicies of K itself. Thus $K \subset \underset{\sim}{K}$ and elements of $\underset{\sim}{K} - K$ are called singular or degenerate simplices. The inclusion $K \hookrightarrow \underset{\sim}{K}$ induces an isomorphism on homology. Further, for each n-simplex Δ_n^{α} of K, $\underset{\sim}{\Delta}_n^{\alpha} \subset \underset{\sim}{K}$ and $\underset{\sim}{\Delta}_n^{\alpha}$ is Kan. Of course, $\underset{\sim}{K} = \bigcup_{\alpha} \underset{\sim}{\Delta}_n^{\alpha}$.

We now consider the Δ-set W. For each i-simplex $\tau: \Delta_i \to W$ of W, we have the induced fibration

$$\begin{array}{ccc} E(\tau) & \xrightarrow{\ \tau^*\ } & E \\ {\scriptstyle \tau(\pi)}\downarrow & & \downarrow{\scriptstyle \pi} \\ \Delta_i & \xrightarrow{\ \tau\ } & W \end{array}$$

Thus we can define the Δ-sets $\tilde{A}_1(\pi)$, $\tilde{A}_2(\pi)$, $\tilde{\rho}_1, \tilde{\rho}_2$ and, for a crossection $\tilde{\psi}$ of $\tilde{A}_2(\pi)$ over $\underset{\sim}{W}$, $\tilde{A}_{\psi}(\pi)$ and $\tilde{\rho}_{\psi}$ exactly as we defined $A_1(\pi), A_2(\pi), \rho_1, \rho_2$, $A_{\psi}(\pi)$ and ρ_{ψ} by simply replacing in all definitions $\pi^{-1}(\Delta_{\tau})$ by $E(\tau)$. The proof that $\tilde{\rho}_1, \tilde{\rho}_2, \tilde{\rho}_{\psi}$ are Kan goes through exactly as the proofs for $\rho_1, \rho_2, \rho_{\psi}$.

Further $A_i(\pi) = \widetilde{A}_i(\pi)|W$, $A_\psi(\pi) = \widetilde{A}_\psi(\pi)|W$, $i = 1, 2$, if $\widetilde{\psi}$ is an extension of ψ.

<u>Proposition 5.3.0.</u> Given $\psi \in \Gamma(A_i(\pi))$ there is a "natural" extension to $\widetilde{\psi} \in \Gamma(\widetilde{A}_i(\pi))$ such that $\widetilde{\psi}|W = \psi$.

<u>Proof.</u> For an ordered simplicial map $\tau: A_i \to K$ we have the degree of degeneracy $= i - \dim \tau(\Delta_i) =$ order τ. Each map of order m factors as

$$\Delta_i \xrightarrow{\tau_{m-1}} \Delta_{i-1} \to \cdots \to \Delta_{i-m} \xrightarrow{\tau_0} K$$

where each a_i is one of the standard collapsing maps and τ_0 is of order 0, i.e., τ_0 is a simplex of K. Then what we have to construct is the following.

Given a crossection of $\widetilde{A}_i(\pi)$ over all simplices of order less than $m > 0$ we wish to extend it to a crossection over simplices of order m. We thus reduce the situation to the following. We are given a diagram and a map $(\lambda, \widetilde{\lambda}): M \to (E(\tau), \tau^*(\mathfrak{e}))$ of type $(1, i-1)$ over $\tau(\pi)$

$$\begin{array}{ccccc}
 & & M & & \\
 & & \downarrow{\scriptstyle \lambda} & & \\
E(\tau a) & \xrightarrow{a^*} & E(\tau) & \longrightarrow & E \\
{\scriptstyle \tau a(\pi)}\downarrow & & \downarrow{\scriptstyle \tau(\pi)} & & \downarrow \\
\Delta_i & \xrightarrow{a} & \Delta_{i-1} & \xrightarrow{\tau} & W
\end{array}$$

and we wish to find $(\lambda', \widetilde{\lambda}'): M' \to (E(\tau a), (\tau a)^*(\mathfrak{e})$ of type $(1, i)$. Let $M', \alpha, \hat{a}$ be the pullback

$$\begin{array}{ccc}
M & \xrightarrow{\hat{a}} & M \\
{\scriptstyle \alpha}\downarrow & & \downarrow{\scriptstyle \tau(\pi)\lambda} \\
\Delta_i & \xrightarrow{a} & \Delta_{i-1}
\end{array}$$

Then α lifts uniquely to $\lambda': M' \to E(\tau a)$ and we have a commutative diagram

$$\begin{array}{ccc} M' & \xrightarrow{\hat{a}} & M \\ {\scriptstyle \lambda'}\downarrow & & \downarrow{\scriptstyle \lambda} \\ E(\tau a) & \xrightarrow{a} & E(\tau) \end{array}$$

We need to show that λ' is an $(i+3)$-ad map of degree one and λ' extends to $(\lambda', \tilde{\lambda}'): M' \to (E(\tau a), (\tau a)^*(\varepsilon))$, a map of type $(1, i)$ over $\tau a(\pi)$.

We will prove only that M' is an $(i+3)$-ad manifold, leaving the rest to the reader. Since all the constructions are natural, this shows that we can extend an element of $\Gamma(A_i(\pi))$ to an element of $\Gamma(\tilde{A}_i(\pi))$. The map a can be defined as follows: Let Δ_{i-1} have vertices $e_0, \ldots, e_{i-1}$. Then the simplicial complex $I \times \Delta_{i-1}$ has vertices $e_0, \ldots, e_{i-1}, e'_0, \ldots, e'_{i-1}$. Let $0 \leq j \leq i-1$. We have an embedding $b_j: \Delta_i \to I \times \Delta_{i-1}$ by extending the vertex map

$$b_j(e_s) = e_s \;, \; 0 \leq s \leq j \,,$$

$$b_j(e_j) = e'_j \;,$$

$$b_j(e_s) = e_{s-1} \;, \; j < s \,.$$

Thus the map $a = pb_j$ for some j, where $p: I \times \Delta_{i-1} \to \Delta_{i-1}$.

Thus for the appropriate function $c_j: \Delta_{i-1} \to [0, 1]$,

$$c_j^{-1}(0) = \partial_j \Delta_i$$

$$c_j^{-1}(1) = e_j$$

$$b_j(\Delta_i) = \{(t, x) \in I \times \Delta_{i-1} \mid c_j(x) \leq t\} \,.$$

Thus $b_j(\Delta_i) \subset I \times \Delta_{i-1}$ is the lower component of $I \times \Delta_{i-1}$ determined by a crossection $c_j \times \mathrm{id}$ of $I \times \Delta_{i-1} \to \Delta_{i-1}$.

We now consider the composite $c_j' = c_j\tau(\pi): M \to [0, 1]$. It follows easily that we can identify M' with the component of $I \times M$ determined by the cross-section $c_j' \times \mathrm{id}$. For M an $(i+2)$-ad manifold, M' has the natural structure of an $(i+3)$-ad manifold and the map λ' is an $(i+3)$-ad map.

This completes the proof of 5.3.0.

For $\psi \in \Gamma(A_i(\pi))$ we will now denote $\widetilde{\psi} = c(\psi)$. Then we have

Proposition 5.3.1. For each $\psi \in \Gamma(A_2(\pi))$ there is a natural embedding $\iota: \Gamma(A_\psi(\pi)) \to \Gamma(\widetilde{A}_\psi(\pi))$ such that $\iota\lambda | W = \lambda$.

Proof. The proof is exactly as in 5.3.0.

For $\gamma \in \Gamma(A_\psi(\pi))$ we denote $\iota(\gamma) = \hat{\gamma}$. The same argument yields

Proposition 5.3.2. Let X be a subcomplex of W. A crossection γ' of $\rho_\psi: A_\psi(\pi) \to W$ over X extends to W if and only if the crossection $\hat{\gamma}'$ of $\widetilde{\rho}_\psi: \widetilde{A}_\psi(\pi) \to \underset{\sim}{W}$ over X extends to $\underset{\sim}{W}$.

W has a natural system of degeneracies and for each simplex Δ_i^α of $\underset{\sim}{W}$ the subcomplex $\underset{\sim}{\Delta}_i^\alpha$ generated by all faces and degeneracies of Δ_i^α is Kan. (This notation agrees with previous notation when $\Delta_i^\alpha \subset W$.) Since ρ_ψ is Kan, $(\widetilde{\rho}_\psi)^{-1}(\underset{\sim}{\Delta}_i^\alpha)$ is a Kan Δ-set and thus its homotopy groups are well-defined. Since $\underset{\sim}{W}$ is connected and simply connected $(\widetilde{\rho}_\psi)^{-1}(\Delta_i^\alpha) \sim (\widetilde{\rho}_\psi)^{-1}(\Delta_j^\alpha)$ and the homotopy equivalence is itself determined up to homotopy. We can then define $\pi_i(\rho) = \pi_i(\rho_\psi) = \pi_{i-1}((\widetilde{\rho}_\psi)^{-1}(\Delta_i))$ for any simplex Δ_i^α of $\underset{\sim}{W}$. We have ignored base-points here. We shall see that for d = formal dimension of $\pi^{-1}(p) \geq 6$, $(\rho_\psi)^{-1}(\Delta_i^\alpha)$ is a loop space and hence basepoints do not matter. Now the usual obstruction theory can be applied to yield, along with 5.3.2, the following corollary.

Corollary 5.3.3. Let $S \subset W$ be a subcomplex. Let ${}^kW$ be the k-th skeleton of W. Suppose γ is a crossection of ρ_ψ over $S \cup {}^kW$. Then γ' defines an element $\underline{\gamma}'$ of $H^{k+1}(W, S, \pi_{k+1}(\rho))$ and $\underline{\gamma}' = 0$ if and only if $\gamma' | S \cup {}^{k-1}W$ extends over $S \cup {}^{k+1}W$.

We now complete the proof of 5.0.1 by identifying the group $\pi_k(\rho)$. This is in fact essentially done by Wall [18], and the results are as follows.

Assume $A_\psi(\pi) \neq \emptyset$. Let $\pi_1 = \pi_1(\pi^{-1}(p))$, $\pi_1^+ = \pi_1(\pi_+^{-1}(p))$, where $\pi: E \to W$, p a vertex of W. Since $(\pi^{-1}(p), \pi_+^{-1}(p))$ is given a P.D. structure, there is defined a Stiefel-Whitney class $w: \pi_1 \to Z_2$ and $w_+: \pi_1^+ \to Z_2$, where $w_+ = w\eta$ and $\eta: \pi_1 \to \pi_1^+$ is induced by the inclusion $\pi_+^{-1}(p) \hookrightarrow \pi^{-1}(p)$. Here π_1, π_1^+ denote the fundamental groupoids. Then

Case I: $\quad \pi_k(\rho) = L_{k+d}(\pi_1, w)$

Case II: $\quad \pi_k(\rho) = L_{k+d}(\eta)$,

where L_{-+-} are the surgery groups of Wall [18].

This completes the proof of 5.0.1.

Remark 1. Utilizing the notation of 5.3.3, if $S \neq \emptyset$, then $A_\psi(\pi) \neq \emptyset$ and γ' always extends over $S \cup {}^0W$. $A_\psi(\pi) = \emptyset$ if and only if $\psi(p)$ has a non-zero surgery obstruction for one and hence every vertex p of W. This obstruction determines an element of

$$\pi_0(\rho) = \begin{cases} L_d(\pi_1, w) \\ L_d(\eta) \end{cases}$$

by convention, which we can identify with $H^0(W, \pi_0(\rho))$. With this convention the theorem as stated is correct without assuming $A_\psi(\pi) \neq \emptyset$.

<u>Remark 2.</u> Suppose we have $\partial_F W = \partial W$ and $\psi \in \Gamma A_2(\pi)$. Then the amalgamation yields a map $\hat{\psi}: M(\psi) \to (E, \mathcal{E})$ of type $(1,0)$. Let $\alpha: \pi_1(E_+) \to \pi_1(E)$ be the map induced by inclusion. Then $\hat{\psi}$ has a well-defined surgery obstruction $s(\hat{\psi})$ in $L_{m+d}(\alpha)$ where m = dimension of W.

On the other hand, we have a canonical crossection γ' of $A_\psi(\pi)$ over $\partial W = \partial_F W$. Suppose γ' extends an ${}^{m-1}W$ to γ^*. Then γ^* determines an obstruction $\hat{\gamma}^*$ to extending the crossection to all of W, where $\hat{\gamma}^* \in H^m(W, \partial W; \pi_m(\rho)) = \pi_m(\rho) = L_{m+d}(\eta)$. Since we have a commutative diagram

$$\begin{array}{ccc} \pi_1(L_+) & \xrightarrow{\ \eta\ } & \pi_1(L) \\ \downarrow & & \downarrow \\ \pi_1(E_+) & \xrightarrow{\ \alpha\ } & \pi_1(E) \end{array}$$

we have a natural morphism $z: L_{n+d}(\eta) \to L_{n+d}(\alpha)$. It is not difficult to show that $z(\hat{\gamma}^*) = s(\hat{\psi})$. Thus when $L_{n+d}(\alpha) \cong L_{n+d}(\eta)$, $z(\gamma^*)$ is determined by $\hat{\psi}$ and is independent of the particular extension γ^* we have selected. This is a very special property of this fibration and implies something about the fiber of ρ. If $W = S^m$ and we know $s(\hat{\psi}) = 0$, our situation with E being associated to $f: V \to W$ and ψ the induced crossection of $A_2(\pi)$, then there is only a single obstruction to extending γ' lying in $L_d(\eta)$, which agrees with 4.1 and 4.2.

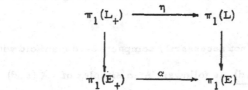

Section 4.

In this section we are going to refine 5.0.1 by proving that the fibration ρ_ψ is induced from a universal object. In the process we will construct Kan Δ sets whose homotopy groups are the surgery obstruction groups. This construction is a refinement of one due to Quinn [22].

For the purpose of this section, a space X will be assumed to come equipped with a map $\pi_1(X) \xrightarrow{w} Z_2$, where π_1 is the fundamental groupoid. If X is a P.D. space, w will be assumed to be the first Stiefel-Whitney class. A map $f: X \to Y$ of spaces X and Y will be assumed to commute with w, i.e., $f^* w_Y = w_X$.

Now let K be a (not necessarily compact) 2-ad manifold with $\iota : K(0) \hookrightarrow K$. We define a Δ-set $\underline{\mathcal{L}(\iota, d)}$ as follows: An n simplex of $\mathcal{L}(\iota, d)$ is a map $(\lambda, \tilde{\lambda}): X \to (Y, \varepsilon)$ over $f: Y \to \Delta_n \times K$ of type $(1, n+1)$, where X is an $(n+4)$-ad manifold, Y an $(n+4)$-ad P.D. space with an s-structure, each of formal dimension $n+d$, and Δ_n is the standard n-simplex. We further assume $\lambda(n+2): X(n+2) \to Y(n+2)$ is an s-equivalence of $(n+3)$-ads.

We regard $\mathcal{L}(\iota, d)$ as a Δ-set by taking $\partial_i(\lambda, \tilde{\lambda}) = (\lambda(i), \tilde{\lambda}(i), f(i))$, $0 \leq i \leq n$, where we identify $\partial_i \Delta_n = \Delta_{n-1}$ via the unique order preserving isomorphism.

Convention: We permit Y or $Y(i)$, thus necessarily X or $X(i)$, to be $\emptyset$. We also permit $K(0) = \emptyset$. In this case $X(n+1), Y(n+1)$ are necessarily empty, and in this case we write $\mathcal{L}(\iota, d)$ as $\mathcal{L}(K, d)$. We always assume $K \neq \emptyset$.

Proposition 5.4.0. $\mathcal{L}(\iota, d)$ is a Kan Δ-set.

Proof. This follows by the argument of 5.2.2.

In our situation $\pi: (E, E_+) \to W$; we wish to find a manifold 2-ad K and an s-equivalence $(E, E_+) \to (K, K(0))$, to use in the construction of $\mathcal{L}(\iota, d)$. The technical problem that arises is that we have only assumed a weak homotopy equivalence $j: (Z, Z_+) \to (E, E_+)$ and therefore τ may not exist. We can get around this as follows.

Construct the Hurewicz fibration λ associated with j:

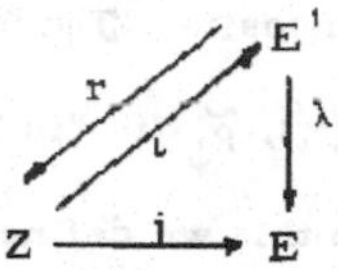

where ι, r are homotopy inverses, λ a weak homotopy equivalence and r, ι, λ maps of 2-ads. Then we can replace the fibration $\pi: (E, E_+) \to W$ by $\pi': (E', E'_+) \to W$, where $E'_+ = E'(0)$, $\pi' = \pi\lambda$ a fibration which has the same homotopy properties and is strongly homotopically equivalent to a finite CW 2-ad Z, Z_+. For any finite CW 2-ad Z, there exists a simple equivalence $(Z, Z_+) \to (K, K(0))$, where K is a (non-compact) 2-ad manifold. Thus we can assume without loss of essential generality that we are given an s-equivalence $u: (E, E_+) \to (K, K(0))$, where K is a 2-ad manifold. For Case I we replace this K by some open extension and let $K(0) = \emptyset$.

Now an n-simplex of $\widetilde{A}_2(\pi)$ is given by a map $(\lambda, \widetilde{\lambda}): X \to (E(\tau), \tau^*(\varepsilon))$ of type $(1, n)$ over $t(\pi): E(\tau) \to \Delta_n$. From this we construct a map of type $(1, n+1)$ $(\lambda, \widetilde{\lambda}): \underline{X} \to (\underline{E(\tau)}, t^*(\varepsilon))$ over $\tau(\pi) \times u\tau: E(\tau) \to \Delta_n \times K$, where in

Case I. $\underline{E(\tau)}$ is the (n+4)-ad which as a space is $E(\tau)$

where $\underline{E(\tau)(i)} = E(\tau)(i) \quad i < n+1$

$\underline{E(\tau)}(n+1) = \emptyset$, $\underline{E(\tau)}(n+2) = E(\tau)(n+1)$.

The exact same formulas relate X and $\underline{X}$ and the maps are the same as the given ones suitably reindexed.

Case II. $\underline{E}(\tau) = E(\tau)(i) \quad 0 \leq i \leq n+1$

$\underline{E}(\tau)(n+2) = \emptyset$.

The same formulas relate X and $\underline{X}$.

The above induces a Δ map $\mathcal{J} : \tilde{A}_2(\pi) \to \mathcal{X}(\iota, d)$. Suppose we are given a crossection $\psi: \underset{\sim}{W} \to \widetilde{A}_2(\pi)$. Then the composite $\mathcal{J}\psi: W \to \mathcal{X}(\iota, d)$ is a Δ map. We will show that in a certain sense $\tilde{A}_\psi(\pi)$, $\tilde{\rho}_\psi$ is "induced" by this map from a universal fibration over $\mathcal{X}(\iota, d)$. To do this we define $E\mathcal{X}(\iota, d)$.

An n-simplex of $E\mathcal{X}(\iota, d)$ is a map $(\lambda, \tilde{\lambda}): X \to (Y, \mathcal{E})$ of type $(1, n+3)$ over $f: Y \to \Delta_n \times K \times I$, where X, Y are of formal dimension $n+d+1$, X an $(n+6)$-ad manifold, Y an $(n+6)$-ad P. D. space such that $\lambda(n+2), \lambda(n+4)$ are s-equivalences of $(n+5)$-ads.

As above, we permit $X(i), X, Y(i), Y$ and $K(0)$, but not K to be empty. $E\mathcal{X}$ is a Δ-set via the usual ∂_i operations and further, given an n-simplex $(\lambda, \tilde{\lambda})$ of $E\mathcal{X}(\iota, d)$ the correspondence $(\lambda, \tilde{\lambda}) \to (\lambda(n+3), \lambda(n+3))$ induces a Δ-map $p: E\mathcal{X}(\iota, d) \to \mathcal{X}(\iota, d)$

<u>Proposition 5.4.0.</u> p is a Kan fibration. For $d \geq 5$, $E\mathcal{X}$ is contractible. Further, for $j+d \geq 6$, $\pi_j(\mathcal{X}(\iota, d)) = L_{j+d}(\iota^*)$, $\iota^*: \pi_1(K(0)) \to \pi_1(K)$ and the basepoint of $\mathcal{X}$ is taken to be the $\emptyset$ object. (When $K(0) = \emptyset$, $L_{j+d}(\iota^*) = L_{j+d}(\pi_1(K), w)$.)

<u>Proof.</u> That $E\mathcal{X}$, $\mathcal{X}$, and p are Kan follows from the argument of 5.2.2. That $E\mathcal{X}$ is contractible and $\mathcal{X}$ has the stated homotopy groups is a consequence of the main theorem of Wall [18].

<u>Corollary 5.4.1.</u> For $d \geq 6$, $\mathcal{X}(\iota, d)$ is the loop space of $\mathcal{X}(\iota, d-1)$.

<u>Proof.</u> $\mathcal{L}(\iota, d)$ can be identified with $p^{-1}(\underset{\sim}{\emptyset})$, $p: E\mathcal{L}(\iota, d-1) \to \mathcal{L}(\iota, d-1)$ and $\underset{\sim}{\emptyset}$ is the Δ-set whose simplices are empty opjects. (One can use this fact to show that $E\mathcal{L}$ is contractible.)

Now the map $\mathcal{J}\psi: \underset{\sim}{W} \to \mathcal{L}(\iota, d)$ is covered in a natural way by $(\mathcal{J}\psi)^*$, and there is a commutative diagram

$$\begin{array}{ccc} \widetilde{A}_\psi(\pi) & \xrightarrow{(\mathcal{J}\psi)^*} & E\mathcal{L}(\iota, d) \\ \Big\downarrow{\scriptstyle \widetilde{p}_\psi} & & \Big\downarrow{\scriptstyle p} \\ W & \xrightarrow{\mathcal{J}\psi} & \mathcal{L}(\iota, d) \end{array}$$

The induced map on $\pi_j(\widetilde{p}_\psi) \to \pi_j(p)$ is just the natural map $L_{j+d}(\eta) \to L_{j+d}(\iota^*)$, where we have the commutative diagram

$$\begin{array}{ccc} \pi_1(\pi_+^{-1}(p)) & \xrightarrow{\eta} & \pi_1(\pi^{-1}(p)) \\ \Big\downarrow & & \Big\downarrow \\ \pi_1(E_+) & \xrightarrow{\iota} & \pi_1(E) \end{array}$$

Now let $\mathcal{L}^0(\iota, d)$ be the sub Δ-set of $\mathcal{L}(\iota, d)$ consisting of those simplices $(\lambda, \mathfrak{X})$ for which λ is an s-equivalence of (n+4)-ads. Then $\mathcal{L}^0(\iota, d)$ is a Kan Δ-set and is contractible.

Let $W' \subset W$ be a subcomplex, ψ a crossection of $A_2(\pi)$ such that $\psi | W' \in A_1(\pi)$. Then $\mathcal{J}\psi(W') \subset \mathcal{L}^0(\iota, d)$.

<u>Theorem 5.4.2.</u> If ψ is homotopic rel W' to a crossection of $A_1(\pi)$, then $\mathcal{J}\psi: W \to \mathcal{L}(\iota, d)$ is homotopic modulo W' to $\alpha: W \to \mathcal{L}^0(\iota, d)$. That is, the map $\mathcal{J}\psi: W/W' \to \mathcal{L}(\iota, d)/\mathcal{L}^0(\iota, d) \sim \mathcal{L}(\iota, d)$ is homotopically trivial. If $L_{j+d}(\eta) \approx L_{j+d}(\iota^*)$, all j, (for example, under assumptions a-h), then the two conditions are equivalent.

Proof. If ρ_ψ has a crossection so does $\tilde{\rho}_\psi$. Thus $\mathcal{J}\psi$ factors through a map into a contractible space and is homotopically trivial. If $\pi_j(\tilde{\rho}_\psi) \approx \pi_j(p)$, then, up to homotopy type, $\tilde{A}_\psi(\pi)$ is the bundle induced via $\mathcal{J}\psi$ from p. Since p has a crossection over $\mathcal{L}^0(\iota, d)$, the result follows.

Remark I. When η and ι^* can be identified, ρ_ψ up to homotopy is just the pull-back of p and $\mathcal{J}\psi$, and the fiber of both up to homotopy is $\mathcal{L}(\iota, d\text{-}1)$. Using a more complicated construction of $\mathcal{L}$, and $E\mathcal{L}$ via restricted objects as in [18,30], one can get $\tilde{A}_\psi(\pi)$ as the actual induced bundle of a map of the form $\mathcal{J}\psi$.

Remark II. By a slight refinement of the main theorem of Wall [18,30], one can prove that $\mathcal{L}(\iota, d)$ up to homotopy type depends only on the algebraic data $\iota_*: \pi_1(K(0)) \to \pi_1(K)$ (at least for $d \geq 6$). In fact, if $g: K \to M$ is a 2-ad manifold map, then g induces a map $\hat{g}: \mathcal{L}(\iota_K, d) \to \mathcal{L}(\iota_M, d)$ which depends only on $g_*: (\pi_1(K), \pi_1(K(0)) \to (\pi_1(M), \pi_1(M(0))$. When K is a single point with $K(0) = \emptyset$, $\mathcal{L}(K, d) = \Omega^d(F/PL)$ -- a space about which a lot is known [29]. For more general ι, little is known about the spaces $\mathcal{L}(\iota, d)$, although they clearly have some striking homotopy theoretic properties (see remarks, Section 3). For example, using amalgamation, it is simple to construct a homomorphism

$$\Omega_n(L(K, d)) \to \pi_n(L(K, d))/2\pi_n(L(K, d))$$

which splits the natural morphism $\pi_n/2\pi_n \to \Omega_n$, where Ω_n denotes unoriented bordism. This suggests that $L(K, d)_{(2)}$ is a product of Eilenberg-MacLane spaces. When K is a point this is a well-known theorem of Sullivan [29].

Section 5. Some Applications.

We have rather thoughly analyzed the problem of deforming $f: V^{m+d} \to W^m$ to a block bundle projection. We now assume f is a block bundle projection and attempt to apply the results of earlier chapters to conclude that, under some additional hypothesis, it is homotopic to a bundle projection. We assume W is connected. We also assume $d \geq 5$ and in Case II, $d \geq 6$.

Let $(F, \partial F) = (f^{-1}(p), f^{-1}(p) \cap \partial_+ V)$ for some vertex p. Then F is a 2-ad manifold. By the s-cobordism theorem, for each simplex Δ_i of W $f^{-1}(\Delta_i) \underset{h}{\simeq} \Delta_i \times F$ as an (i+3)-ad manifold. In case I, $\pi h(i+1): f^{-1}(\Delta_i) \cap \partial_+ V \to \Delta_i$ is the same as f.

Thus V is a block bundle over W with fiber F and $\partial_+ V$ is a sub block bundle which, in Case I, is a fiber bundle [25].

Let $\widetilde{\mathcal{A}}(F)$ be the Δ-group whose k-simplices are (k+3)-ad homeomorphisms $\lambda : \Delta_k \times F \simeq \Delta_k \times F$ (pℓ or smooth depending on our category). We set $\partial_i \lambda = \lambda(i)$. Let $\widetilde{\mathcal{A}}(F, \partial F)$ be the subgroup such that $\lambda(k+1): \Delta_k \times \partial F \to \Delta_k \times \partial F$ commutes with projection onto Δ_k. Let $\mathcal{A}(F) \subset \widetilde{\mathcal{A}}(F, \partial F)$ be the subgroup such that λ commutes with projection onto Δ_k.

These Δ-groups all have classifying spaces

$$B\mathcal{A}(F) \hookrightarrow B\widetilde{\mathcal{A}}(F, \partial F) \hookrightarrow B\widetilde{\mathcal{A}}(F).$$

By the results of Rourke-Sanderson [28], f determines a unique homotopy class of Δ-maps

$$\hat{f}: W \to B\widetilde{\mathcal{A}}(F) \quad \text{in Case II},$$

$$\hat{f}: W \to B\widetilde{\mathcal{A}}(F, \partial F) \quad \text{in Case I},$$

such that f is homotopic (through block bundle projections) to a fiber bundle projection modulo $\partial_F V$ in Case II and to $\partial_F V \cup \partial_+ V$ in Case I, if and only if f is homotopic modulo $\partial_F W$ to a map which factors through $\hat{\hat{f}}: W \to B\mathcal{A}(F)$.

We can now apply the results of the earlier chapters. By 3.2, 3.3 and the remark following 3.3, we have conditions for $\pi_i(\mathcal{J}_1)$ and $\pi_i(\mathcal{J}_2)$ to vanish in the $p\ell$ category, where

$$\mathcal{J}_1 : B\mathfrak{a}(F) \to B\tilde{\mathfrak{a}}(F, \partial F),$$
$$\mathcal{J}_2 : B\mathfrak{a}(F) \to B\tilde{\mathfrak{a}}(F) .$$

We write these conditions below for future reference and refer to them as

Conditions 5.5:

(a) $\pi_i(\mathcal{J}_1) = 0$ for $i \leq k$ if $\pi_j(F) = 0$ for $j \leq s$, $d \geq 7$ and any one of the following conditions hold.

1a. $k \leq \inf(2s-2, s+4)$ and $\pi_1(\partial F) = 0$,

2a. $k \leq \inf(2s-2, s+4)$ and $s \leq d-4$,

3a. $k \leq 2s-1$, $s = 2, 4, 5$, $s \leq d-4$, $\pi_1(\partial F) = 0$,

4a. $k \leq 2s-1$, $s = 2, 4, 5$, $s \leq d-5$,

5a. $k \leq 5$, $s = 3$, $\pi_1(\partial F) = 0$ and $\tau({}^4F)$ is trivial,

6a. $k \leq 5$, $s = 3$, $d \geq 8$, $\tau({}^4F)$ is trivial.

(Recall that ${}^4F$ is a neighborhood of the 4-skeleton of F.)

(b) Let $\partial F \neq 0$, $d \geq 8$; then $\pi_i(\mathcal{J}_2) = 0$ for $i \leq k$ if $\pi_j(F) = \pi_j(\partial F) = 0$ for $j \leq s$ and any one of the following conditions hold.

1b. $k \leq \inf(2s-2, s+4)$,

2b. $k \leq 2s-1$, $s = 2, 4, 5$, $s \leq d-5$,

3b. $k \leq 5$, $s = 3$, $d \geq 9$ and $\tau({}^4F)$ is trivial.

From this we conclude

Theorem 5.5.0. Let f be a block bundle projection. Suppose $H^p(W, \partial_F W)$ vanishes for any coefficients and $p > k$. Suppose F satisfies any one of 1a-6a in Case I or $(F, \partial F)$ satisfies any one of 1b-3b in Case II. Then f is homotopic, modulo $\partial_F V \cup \partial_+ V$ in Case I and modulo $\partial_F V$ in Case II (through block bundle projections) to a bundle projection.

For any $r > 0$, consider the maps

$$\mathcal{J}_1^r : B\mathcal{A}(F \times T^r) \to B\tilde{\mathcal{A}}(F \times T^r, \partial F \times T^r)$$

and

$$\mathcal{J}_1^r : B\mathcal{A}(F \times T^r) \to B\tilde{\mathcal{A}}(F \times T^r),$$

where T^r is the r-torus $\prod^r S^1$. We have seen (Ch. 3) that, for sufficiently large r depending on k, the maps $\pi_i(\mathcal{J}_1) \to \pi_i(\mathcal{J}_1^r)$ and $\pi_i(\mathcal{J}_2) \to \pi_i(\mathcal{J}_2^r)$ are zero for $i \leq k$. Thus we have

Theorem 5.5.1. Let f be a block bundle projection. Then for sufficiently large r (depending on the dimension of W), $f^r: V \times T^r \to W$ is homotopic through block bundle projections to a bundle projection, where $f^r = f_p$. (This is true in both the $p\ell$ and smooth categories).

Now using the remark at the end of Section 2, the remark following 5.0.4 and 5.0.3, we can conclude

Theorem 5.5.2. Let $f: V \to W$ satisfy assumptions a-g. Let K be a compact, connected manifold with connected, non-empty boundary such that $\pi_1(\partial K) \approx \pi_1(K)$. Then for sufficiently large r, $f_K^r : V \times K \times T^r \to W$ is homotopic to a bundle projection.

None of the constructions of Sections 2-5 depend on W being compact. However, the interesting geometric conclusions depend on 5.1.0 which does involve the compactness of V. In the case where V is non-compact, we can get certain results by thinking of V as the union of compact sub-manifolds. In fact, in this situation, we can often avoid the troublesome assumption h. Thus the situation is in a sense more tractible for open manifolds.

We now consider a <u>proper</u> map $f: V \to W$. All deformations are to remain in the category of proper maps, and we henceforth consider only maps in this category; homotopy in this category means proper homotopy.

Consider a non-compact 3-ad manifold $(W, \partial_F W, \partial_- W)$ such that there exists a compact 4-ad manifold $(\overline{W}, \partial_F \overline{W}, \partial_- \overline{W}, \partial_1 \overline{W})$ with

$$W = \overline{W} - \partial_1 \overline{W},$$
$$\partial_F W = \partial_F \overline{W} - \partial_1 \overline{W},$$
$$\partial_- W = \partial_- \overline{W} - \partial_1 \overline{W}.$$

Then we have

<u>Theorem 5.5.3.</u> Suppose $\pi_1(\partial_1 \overline{W}) = 0$, $\partial_1 \overline{W}$ connected and $f: V \to W$ satisfies assumptions a-g. Then all the previous theorems are true in this category.

<u>Proof.</u> The only non trivial thing to prove is that if $A_\psi(\pi)$ has a cross-section, f is (properly) homotopic to a block bundle projection.

We consider the following decomposition of W

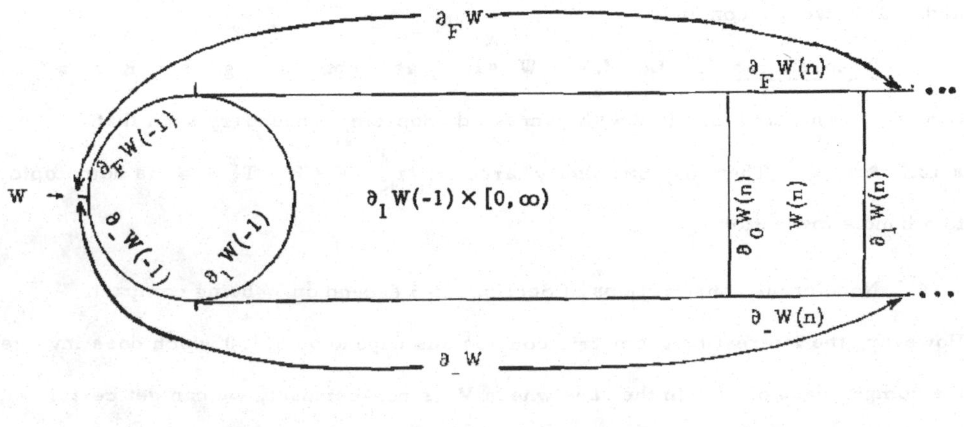

Here the 4-ad manifold $(W(-1), \partial_F W(-1), \partial_- W(-1), \partial_1 W(-1))$ is a smaller copy of $\overline{W}$ inside of W and $W = W(-1) \cup \partial_1 W(-1) \times [0, \infty)$.

We also have the 5-ad manifolds

$$(W(n), \partial_0 W(n), \partial_F W(n), \partial_- W(n), \partial_1 W(n))$$
$$= (\partial_1 W(-1) \times [n, n+1], \partial_1 W(-1) \times (n), \partial_F W \cap \partial_1 W(-1) \times [n, n+1],$$
$$\partial_- W \cap \partial_1 W(-1) \times [n, n+1], \partial_1 W(-1) \times (n+1)) .$$

Let $\pi: (E, E_+) \to W$ be the Hurewicz fibration associated with f. We then have the induced fibrations $E(-1), E(n)$ over $W(-1)$ and $W(n)$, where $E(-1), E(n)$ are 5-ad and 6-ad P.D. spaces, respectively. We consider the 4-ad s-equivalence $r: V \to E$ with $\pi r = f$. Then a standard surgery argument (e.g. proof of 11.3 in [18], although the hypotheses are different) allows us to deform r to $\bar{r}$ so that $\pi\bar{r}$ is transverse to each $W(n)$ and $V(n) = r^{-1}\pi^{-1}(W(n)$, $\bar{r}|V(n): V(n) \to E(n)$ is an s-equivalence of 6-ads (5-ads if n= -1). It follows that a crossection of $A_\psi(\pi)$ yields a crossection of $A_\psi(\pi)|W(-1)$ and hence a homotopy of $\bar{r}^1$ to r^0 such that $\pi r^0|V(-1)$ is a block bundle projection.

Now consider $A_\psi(\pi)|W(0)$. Since $\pi r^0|V(-1) \cup \partial_F V$ is already a block bundle projection (in Case I, $\pi r^0|V(-1) \cup \partial_F V \cup \partial_+ V$), we have a crossection of $A_\psi(\pi)$ over $\partial_0 W(0) \cup \partial_F W(0)$ and since $H_*(W(0), \partial_0 W(0) \cup \partial_F W(0)) = 0$ we can extend this to a crossection over all of $W(0)$. We thus have a crossection over $W(-1) \cup W(0)$. Thus r^0 can be deformed modulo $V(-1) \cup \partial_F V$ (and $\partial_+ V$ in Case I) to r^1, where $\pi r^1|V(-1) \cup V(0)$ is a block bundle projection. We can continue in this way to deform r^{n+1} to r^{n+2} such that $\pi r^{n+1}|V(-1) \cup V(0) \cup \dots \cup V(n)$ is a block bundle projection modulo $V(-1) \cup \dots \cup V(n) \cup \partial_F V$, with $\pi r^{n+2}|V(-1) \cup V(0) \cup \dots \cup V(n+1)$ a block bundle projection. Piecing these deformations together yields $r^\infty: V \to E$ with πr^∞ a block bundle projection and $f = \pi r$ homotopic to $f' = \pi r^\infty$. Q.E.D.

We now consider a more general situation. We assume $(W, \partial_F W, \partial_- W) = (\bigcup_n W(n), \bigcup_n \partial_F W(n), \bigcup_n \partial_- W(n))$, where $-1 \leq n < \infty$ and each $W(n)$ is a compact 5-ad manifold $(W(n), \partial_0 W(n), \partial_F W(n), \partial_- W(n), \partial_1 W(n))$. We assume $W(n), \partial_0 W(n), \partial_- W(n), \partial_1 W(n)$ are all connected and simply connected and $W(n) \cap W(n+1) = \partial_1 W(n) = \partial_0 W(n+1)$ and $W(i) \cap W(j) = \emptyset$ for $|i-j| \geq 2$. We call such a decomposition of $(W, \partial_F W, \partial_- W)$ a <u>simply connected decomposition.</u>

<u>Theorem 5.5.4.</u> Let $f: V \to W$ be a proper map satisfying assumptions a-g, except that V, W need not be compact. Suppose W has a simply connected decomposition and assume $\pi_1(\partial L) = \pi_1(L)$ and we are in Case II. Then f is properly homotopic to a block bundle projection.

<u>Proof.</u> We proceed exactly as in the proof of the previous theorem, replacing the no-longer true statement that $H^*(W(n), \partial_0 W(n) \cup \partial_F W(n)) = 0$ for all coefficients with the statement that it is true for coefficients $L(\iota^*)$, since $\iota^*: \pi_1(\partial L) \approx \pi_1(L)$ and thus $L(\iota^*) = 0$.

<u>Corollary 5.5.5.</u> Let $f: V \to W$ satisfy the hypothesis of 5.5.4. If $H^p(W)$ vanishes for $p > k$ and all coefficients and if k satisfies any of the conditions 1b, 2b, 3b of 5.5, then f is properly homotopic to a p.ℓ. bundle projection.

<u>Corollary 5.5.6</u> (Splitting Theorem). Let $(L, \partial L)$ be a finite P.D. space with $L, \partial L, (\partial L \neq \emptyset)$ connected, and $0 = \pi_i(L) \approx \pi_i(\partial L)$ $0 \leq i < k \geq 6$ and L of formal dimension ≥ 6. Suppose K is a closed, connected and simply connected manifold of dimension $\leq k+4$. Suppose further that $V \underset{p}{\sim} R^n \times K \times (L, \partial L)$ (where $\underset{p}{\sim}$ denotes properly homotopically equivalent) and $n \geq 1$, $n \neq 2$. Then the map $\pi p: V \to R^n \times K$ is properly homotopic to a p.ℓ. bundle map with fiber a p.ℓ. manifold $(F, \partial F) \sim (L, \partial L)$.

If K is a point, the hypothesis on $\pi_i(L), \pi_i(\partial L)$ can be weakened to $\pi_i(L) \approx \pi_i(\partial L)$ because in this case $V \to R^n$ is homotopic to a block bundle projection, and since R^n is contractible any map of R^n to $B\widetilde{\mathcal{A}}(F)$ must lift to $B\mathcal{A}(F)$. Then we have $V \to R^n$ is homotopic to a fiber bundle projection and since R^n is contractible $V \simeq F \times R^n$, i.e., the bundle must be trivial. For K a point, the argument also works in the smooth category.

6. The structure of $A(M \times S^1)$

Let M^n be a connected compact manifold and S^1 the unit sphere in the complex plane with the usual orientation. Let $S^1_+ = \{z \in S^1 \mid \mathrm{Real}(z) \geq 0\}$ and $S^1_- = \{z \in S^1 \mid \mathrm{Real}(z) \leq 0\}$. Then we have the bundle

$$A(M \times S^1_-) \xrightarrow{i} A(M \times S^1) \xrightarrow{\rho} E'(M \times S^1_+, M \times S^1),$$

where $E'(M \times S^1_+, M \times S^1)$ is the subspace of the space of embeddings consisting of those connected components in the image of $A(M \times S^1)$ by the restriction map ρ.

Now $E(M \times S^1_+, M \times S^1)$ is homotopy equivalent under restriction to $E(M, M \times S^1)$. Thus if we identify S^1_- to $I = [-1, 1]$ by an orientation preserving isomorphism, the above fibration becomes

$$(*) \qquad A(M \times I) \xrightarrow{i} A(M \times S^1) \xrightarrow{\rho} E'(M, M \times S^1).$$

The main result of this section (which will be proved at the end of the section) is that $(*)$ is essentially trivial:

Theorem 1. a) If $M^n = N^{n-1} \times I$, then $(*)$ is trivial.

b) For any M, the pull-back of $(*)$ over any finite dimensional skeleton of $E'(M, M \times S^1)$ is trivial.

Similar considerations give a fibration

$$C(M \times I) \xrightarrow{i} C(M \times S^1) \xrightarrow{\rho} C'(M, M \times S^1),$$

where $C'(M, M \times S^1)$ are the components of $C(M, M \times S^1)$ in the image of ρ. All our results apply to $C(M \times S^1)$ as well, but we will only state them for $A(M \times S^1)$.

In case $\partial M = \emptyset$, we have an action of S^1 on $M \times S^1$ by ident. $\times$ left translation, and hence an action of S^1 on $A(M \times S^1)$ and $E(M, M \times S^1)$ by composition. Choose a basepoint $x \in M$ and let $\overline{A}(M \times S^1)$, resp. $\overline{E}(M, M \times S^1)$, be the subspace which sends $(x, 1)$, resp. x, into $M \times (1)$. Then it is trivial to prove:

Lemma 1. If $\partial M = \emptyset$, $A(M \times S^1) = \overline{A}(M \times S^1) \times S^1$, and

$$E(M, M \times S^1) = \overline{E}(M, M \times S^1) \times S^1.$$

Note that if $\partial M \neq \emptyset$ and we take $x \in \partial M$, then since ∂M is fixed in $A(M \times S^1)$ and $E(M, M \times S^1)$, we have: If $\partial M \neq \emptyset$, $\overline{A}(M \times S^1) = A(M \times S^1)$ and $\overline{E}(M, M \times S^1) = E(M, M \times S^1)$.

For $\overline{A}(M \times S^1)$, resp. $\overline{E}(M, M \times S^1)$, we have a well-defined map $\lambda: \overline{A}(M \times S^1) \to \overline{A}(M \times R)$, resp. $\lambda: \overline{E}(M, M \times S^1) \to \overline{E}(M, M \times R)$, where $\overline{A}(M \times R)$, resp. $\overline{E}(M, M \times R)$, send $(x, 0)$, resp. x, to $M \times (0)$, by lifting to the universal cover. Note that $\overline{A}(M \times R)$ is a deformation restract of $A(M \times R)$, and $\overline{E}(M, M \times R)$ is a deformation retract of $E(M, M \times R)$.

Let $j: R \to \text{Int}\, S^1_+$ be an orientation preserving isomorphism $j(0) = 1$. Then j defines $j: \overline{E}(M, M \times R) \to \overline{E}(M, M \times S^1)$. Then by uniqueness of collars we have:

Lemma 2. $\overline{E}(M, M \times R) \xrightarrow{j} \overline{E}(M, M \times S^1) \xrightarrow{\lambda} \overline{E}(M, M \times R)$ is homotopic to the identity.

Let $E_0(M, M \times R)$ be the component of the inclusion in $E(M, M \times R)$.

Lemma 3. $E_0(M, M \times R)$ is homotopy equivalent to $BA(M \times I)$, the universal base space for $A(M \times I)$.

Proof. Consider the fibration

$$A(M \times [-1,0]) \to E_{M \times (-1)}(M \times [-1, 0], M \times R) \to E_0(M, M \times R),$$

where the middle term is the oriented embeddings of $M \times [-1, 0]$ in $M \times R$ which are the identity on $(\partial M \times [-1, 0]) \cup (M \times (-1))$. But this space is contractible by uniqueness of collars. The lemma follows.

Now let $\overline{E}_0(M, M \times R)$, respe $\overline{E}_0(M, M \times S^1)$, be the component of the inclusion in $\overline{E}(M, M \times R)$, resp. $\overline{E}(M, M \times S^1)$. Then $\overline{E}_0(M, M \times R)$ is a deformation retract of $E_0(M, M \times R)$ and when $\partial M \neq \emptyset$, $\overline{E}_0(M, M \times R) = E_0(M, M \times R)$ and $\overline{E}_0(M, M \times S^1) = E_0(M, M \times S^1)$.

Definition. $\mathcal{N}(M)$ is the homotopy theoretic fibre of $\lambda : \overline{E}_0(M, M \times S^1) \to \overline{E}_0(M, M \times R)$.

By Theorem 1 and Lemmas (2) and (3) we get:

Theorem 2: (i) If $M = N \times I$,

$$\Omega A(M \times S^1) \simeq \Omega A(M \times I) \times A(M \times I) \times \Omega \mathcal{N}(M).$$

(ii) For any M,

$$\pi_i A(M \times S^1) \simeq \pi_i A(M \times I) \oplus \pi_{i-1} A(M \times I) \oplus \pi_i \mathcal{N}(M), \quad i > 0.$$

Finally, we have:

Theorem 3. Let $n \geq 5$. For M^n a smooth manifold, $\mathcal{N}^d(M) \simeq \mathcal{N}^{p\ell}(M) \simeq \mathcal{N}^t(M)$; and for M a PL manifold, $\mathcal{N}^{p\ell}(M) \simeq \mathcal{N}^t(M)$. Hence $\mathcal{N}(M)$ is a topological invariant.

Proof. For simplicity we will do only the Diff-Top case, the other cases are entirely similar. It is enough to show that:

$$(\overline{E}_0^t(M\times S^1_+, M\times S^1), \overline{E}_0^d(M\times S^1_+, M\times S^1)) \xrightarrow{\lambda} (\overline{E}_0^t(M\times S^1_+, M\times R), \overline{E}_0^d(M\times S^1_+, M\times R)),$$

induces isomorphisms on homotopy groups. By Theorem 3.1 of [2], this is equivalent to showing the same for

$$(\overline{Im}_0^t(M\times S^1_+, M\times S^1), \overline{Im}_0^t(M\times S^1_+, M\times S^1)) \xrightarrow{\lambda} (\overline{Im}_0^t(M\times S^1_+, M\times R), \overline{Im}_0^t(M\times S^1_+, M\times R));$$

but this is obvious.

Proof of Theorem 1a: Let $M = N \times I$. By identifying I with S^1_- by an orientation preserving isomorphism as in (*), we have an embedding $I \times I \subset I \times S^1$. Also we may embed S^1 in $\operatorname{Int} I \times I$ with a trivial normal tube to get an embedding $S^1 \times I \subset \operatorname{Int} I \times I$. These induce embeddings $N \times I^2 \to N \times I \times S^1 \to N \times I^2$ or $M \times I \to M \times S^1 \to M \times I$, and homomorphisms $A(M\times I) \xrightarrow{i} A(M\times S^1) \xrightarrow{j} A(M\times I)$ such that ji is homotopic to the identity. These in turn induce maps of the universal base spaces: $BA(M \times I) \xrightarrow{\overline{i}} BA(M \times S^1) \xrightarrow{\overline{j}} BA(M \times I)$ such that $\overline{j}\overline{i}$ is homotopic to the identity. Therefore, the classifying map $\varphi: E^1(M, M\times S^1) \to BA(M\times I)$ of $(*)$ is homotopically trivial since $\varphi \sim \overline{j}\,\overline{i}\,\varphi$ and $\overline{i}\varphi$ is homotopically trivial.

For part b) of Theorem 1 we will need (cf 3.12)

Proposition 4. $\lambda: \overline{A}_0(M\times S^1) \to \overline{A}_0(M\times R)$ has a right homotopy inverse; i.e., there exists $\mu: \overline{A}_0(M\times R) \to \overline{A}_0(M\times S^1)$ such that $\lambda\mu \sim$ identity.

Proof. We have the commutative diagram:

$$\begin{array}{ccc} \overline{A}(M\times S^1) & \xrightarrow{\ \rho\ } & \overline{E}(M, M\times S^1) \\ \lambda\Big\downarrow & & \Big\downarrow\lambda \\ \overline{A}(M\times R) & \xrightarrow{\ \rho\ } & \overline{E}(M, M\times R)\ . \end{array}$$

Since the fibre of $\rho: \overline{A}(M\times R) \to \overline{E}(M, M\times R)$ is $A(M\times R; M\times 0)$, which is contractible, ρ is a homotopy equivalence. Hence it is sufficient to find a map $\nu: \overline{E}_0(M, M\times R) \to \overline{A}_0(M\times S^1)$ with $\lambda\rho\nu \sim$ identity; or by Lemma 2 -- with $\rho\nu \sim j: \overline{E}_0(M, M\times R) \to \overline{E}_0(M, M\times S^1)$.

Now we have a fibration

$$(1) \qquad A(M\times I)\times A(M\times I) \xrightarrow{\ \alpha\ } \overline{A}(M\times I) \xrightarrow{\ \rho\ } \overline{E}_0(M, M\times R)\ .$$

In fact, if we let $I = [-1, 1]$ and identify R with $(-1, 1)$, and consider that ρ factors, $\rho: \overline{A}(M\times I) \to \overline{E}_0(M\times[-\varepsilon, \varepsilon], M\times(-1,1)) \to \overline{E}_0(M, M\times R)$ with the last map a homotopy equivalence; we see that the fibre of ρ is $A(M\times I; M\times[-\varepsilon, \varepsilon])$. But this may be identified with $A(M\times I)\times A(M\times I)$.

Now (1) maps into $(*)$:

$$\begin{array}{ccccc} A(M\times I)\times A(M\times I) & \xrightarrow{\ \alpha\ } & \overline{A}(M\times I) & \xrightarrow{\ \rho\ } & \overline{E}_0(M, M\times R) \\ \Big\downarrow i_1 & & \Big\downarrow i_2 & & \Big\downarrow j \\ A(M\times I) & \xrightarrow{\ i\ } & A(M\times S^1) & \xrightarrow{\ \rho\ } & E'(M, M\times S^1) \end{array}$$

where i_2 is obtained by identifying I with S^1_+, j by identifying R with Int S^1_+

as in Lemma 2, and i_1 is defined to make the diagram commutative. In fact, if τ_0 is the isomorphism $\tau_0(x,t) = (x,-t)$ of $M \times I$ and τ is the automorphism of $A(M \times I)$ given by $\tau(h) = \tau_0 h \tau_0^{-1}$, then $i_1 = \alpha \circ \tau \times \tau$. Note that i_1 is a homomorphism and therefore we can extend our map of principal bundles to:

$$\begin{array}{ccccc} & & \overline{E}_0(M, M \times R) & \xrightarrow{\varphi'} & BA(M \times I) \times BA(M \times I) \\ & & \downarrow j & & \downarrow \overline{\iota}_1 \\ \overline{A}_0(M \times S^1) & \xrightarrow{\rho} & E_0(M, M \times S^1) & \xrightarrow{\varphi} & BA(M \times I) \end{array}$$

We need to show $\varphi \circ j = \overline{\iota}_1 \circ \varphi'$ is trivial. Then j will lift to

$$\nu : \overline{E}_0(M, M \times R) \to \rho^{-1}(\overline{E}_0(M, M \times S^1)) \subset \overline{A}(M \times S^1),$$

which can be assumed in $\overline{A}_0(M \times S^1)$ since $A(M \times I)$ is transitive on components of $\rho^{-1}(\overline{E}_0(M, M \times S^1))$.

Since $\alpha : A(M \times I) \times A(M \times I) \to A(M \times I)$ is a homomorphism, it induces $\overline{\alpha} : BA(M \times I) \times BA(M \times I) \to BA(M \times I)$. It is clear that $\overline{\alpha}$ makes $BA(M \times I)$ into a homotopy associative H-space with homotopy unit. We will show below that it has a homotopy inverse as well. Assume this for a moment. Let $k_1 = \pi_1 \circ \varphi'$ and $k_2 = \pi_2 \circ \varphi'$. Then $\overline{\alpha} \circ k_1 \times k_2$ is trivial. On the other hand, $\overline{i}_1 \circ \varphi' = \overline{i}_1 \circ k_1 \times k_2 = \overline{\alpha} \circ \overline{\tau} k_1 \times \overline{\tau} k_2$. Now $\alpha \circ \tau \times \tau = \tau \circ \alpha \circ \gamma$, $\gamma(a,b) = (b,a)$, $a, b \in A(M \times I)$. Hence $\overline{\alpha} \circ \overline{\tau} k_1 \times \overline{\tau} k_2 = \overline{\tau} \circ \overline{\alpha} \circ k_2 \times k_1$. Since $[\ \ , BA(M \times I)]$ forms a group, $[k_1][k_2] = e$ implies $[k_2][k_1] = e$, and $\overline{\alpha} \circ k_2 \times k_1$ is trivial. Since $\overline{\tau}$ is a homotopy equivalence, $\overline{i}_1 \circ \varphi' = \overline{\tau} \circ \overline{\alpha} \circ k_2 \times k_1$ is trivial, and j lifts.

It remains to show $BA(M \times I)$ has a homotopy inverse.[1]

---

[1] Alternately, one may show $BA(M \times I)$ is a loop space by May's recognition principle.

From the map of fibrations

$$\begin{array}{ccccc} A(M \times I) \times A(M \times I) & \xrightarrow{\alpha} & A(M \times I) & \longrightarrow & E_0(M, M \times R) \\ \downarrow{\scriptstyle \pi_1} & & \downarrow{\scriptstyle \rho} & & \| \\ A(M \times I) & \longrightarrow & E_{M \times (-1)}(M \times I, M \times R) & \longrightarrow & E_0(M, M \times R) \end{array}$$

we get a commutative diagram

$$\begin{array}{ccccc} E_0(M, M \times R) & \xrightarrow{\varphi'} & BA(M \times I) \times BA(M \times I) & \xrightarrow{\overline{\alpha}} & BA(M \times I) \\ \| & & \downarrow{\scriptstyle \pi_1} & & \\ E_0(M, M \times R) & \xrightarrow{\varphi''} & BA(M \times I) & & \end{array}$$

where the classifying map φ'' is a homotopy equivalence. Consequently, $\psi = \varphi' \circ (\varphi'')^{-1} : BA(M \times I) \to BA(M \times I) \times BA(M \times I)$ satisfies $\pi_1 \circ \psi \sim$ identity and $\overline{\alpha} \circ \psi$ is trivial. It follows that $\pi_2 \circ \psi$ defines a right homotopy inverse. Similarly, one may show the existence of a left homotopy inverse and hence a two-sided homotopy inverse using the homotopy associativity.

<u>Remark.</u> Since λ commutes with the action of $A(M \bmod (x))$, $\lambda : \overline{A}(M \times S^1) \to \overline{A}(M \times R)$ has a right homotopy inverse.

<u>Corollary.</u> If $A(M \times I) \to j^* A(M \times S^1) \xrightarrow{\rho} \overline{E}_0(M, M \times R)$ is the fibration induced from (*) by $j : \overline{E}_0(M, M \times R) \to E'(M, M \times S^1)$, then the fibration is trivial.

To prove Theorem 1(b), it is obviously enough to restrict our attention to

$$A(M \times I) \to A_\rho(M \times S^1) \xrightarrow{\rho} E_0(M, M \times S^1),$$

where $A_\rho(M \times S^1) = \rho^{-1}(E_0(M, M \times S^1))$, since ρ commutes with left action of $A(M \times S^1)$. But then since ρ commutes with the action of S^1 it is sufficient to consider

$$A(M \times I) \to \overline{A}_\rho(M \times S^1) \xrightarrow{\rho} \overline{E}_0(M, M \times S^1).$$

Now let $c(p): \overline{E}_0(M, M \times S^1) \to \overline{E}_0(M, M \times S^1)$ be the lift to the p-fold covering, p a prime. Write $A = \overline{E}_0(M, M \times S^1)$ and $B = \overline{E}_0(M, M \times R)$. Now $c(p): A \to A$ is an inclusion and we have the direct limit: $A \xrightarrow{c(p)} A \xrightarrow{c(p)} A \to \cdots \Rightarrow \overline{A}$

Further, we have the commutative diagram

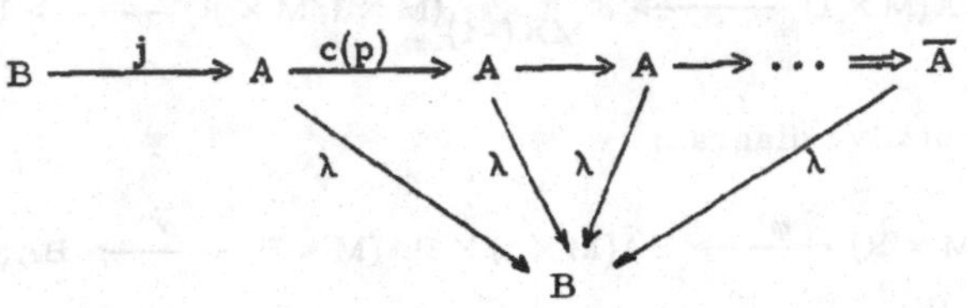

For any $\alpha \in \pi_i(A)$, $c(p)^n(\alpha)$ is in the image from $\pi_i(B)$ if n is sufficiently large. Hence $\pi_i(B) \to \pi_i(\overline{A})$ is onto, and since $\overline{\lambda} \circ j \sim$ identity, $\overline{\lambda}$ is a homotopy equivalence.

Now write $G = \overline{A}_\rho(M \times S^1)$ and $H = A(M \times I)$. $c(p): G \to G$ is also an inclusion and the direct limit $\overline{G}$ exists. Now we may define an inclusion $p: H \to H$ such that

$$\begin{array}{ccc} H & \xrightarrow{p} & H \\ {\scriptstyle i}\downarrow & & \downarrow{\scriptstyle i} \\ G & \xrightarrow{c(p)} & G \end{array}$$

commutes.

The map $p: A(M \times I) \to A(M \times I)$ is obtained by subdividing I into p subintervals, and for $h \in A(M \times I)$, taking a copy of h (reparameterized) on each subinterval. Note that p(h) is homotopic to h^p. It follows that the direct limit $\overline{H}$ of H under p^n is $H_{(p)}$; i.e., H localized away from p. Now p is a homomorphism and $\overline{H}$ is a group acting on $\overline{G}$ with quotient $\overline{A}$; i.e., $\overline{G} \to \overline{A}$ is a fibration and we have a map of fibrations

$$\text{(1)} \qquad \begin{array}{ccccc} H & = & H & \longrightarrow & \overline{H} \\ \downarrow & & \downarrow & & \downarrow \\ j^*G & \to & G & \longrightarrow & \overline{G} \\ \downarrow & & \downarrow & & \downarrow \\ B & \to & A & \longrightarrow & \overline{A} \end{array}$$

By Proposition 4, $j^*G \to B$ is trivial.

Take the diagram (1) and localize the whole diagram away from p:

(2)
$$\begin{array}{ccccc} H_{(p)} & = & H_{(p)} & = & H_{(p)} \\ \downarrow & & \downarrow & & \downarrow \\ (j^*G)_{(p)} & \longrightarrow & G_{(p)} & \longrightarrow & \overline{G}_{(p)} \\ \downarrow & & \downarrow & & \downarrow \\ B_{(p)} & \longrightarrow & A_{(p)} & \longrightarrow & \overline{A}_{(p)} \end{array}$$

This is again a diagram of fibrations; and since $B \to \overline{A}$ and hence $B_{(p)} \to \overline{A}_{(p)}$ is a homotopy equivalence, the map from the first to the third fibration in (2) is an equivalence of fibrations. Hence $\overline{G}_{(p)} \to \overline{A}_{(p)}$ is trivial and the induced fibration $G_{(p)} \to A_{(p)}$ must also be trivial. Thus the classifying map $\varphi: A \to BH$ has the property that $\varphi_{(p)}: A_{(p)} \to BH_{(p)}$ is trivial for each p. Thus Theorem 1(b) follows from

Theorem (Peter May)[1]: Let $f: X \to Y$ be a map of CW complexes, X countable, Y nilpotent, such that $f_{(p)}: X_{(p)} \to Y_{(p)}$ is trivial for each p. Then for any finite-dimensional subcomplex $X^{(k)}$ of X, $f: X^{(k)} \to Y$ is trivial.

---

[1] On duality and completions in homotopy theory -- to appear.

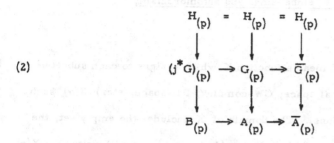

Appendix 1

Kan sets of embeddings and automorphisms

§1. n-ads

By an n+1 -ad we will mean a function X which assigns to each subset α of $\{1, 2, \ldots, n\}$ a set (topological space, CW-complex, PL-space, etc.) $X(\alpha)$ such that X preserves intersections. (The domain of X includes the empty set, the range may or may not.) In particular, if $\alpha = \{1, 2, \ldots, n\}$ we will write $X = X(\alpha)$. Note that it is sufficient to choose X and n-subsets (subspaces, subcomplexes, PL-subspaces, etc.) $X_i = X(\alpha_i)$, $\alpha_i = \{1, 2, \ldots, \hat{i}, \ldots, n\}$, whose intersections are again in the given category. Consequently, we may denote the n+1-ad X by $(X; X_1, \ldots, X_n)$.

Example 1. Let Δ^{n-1} be the standard n-1-simplex in R^n, then $(\Delta^{n-1}, \partial_0\Delta^{n-1}, \ldots, \partial_{n-1}\Delta^{n-1})$ is an n+1-ad.

Example 2. If $(X; X_1, \ldots, X_n)$ is an n+1-ad,

$$(X_i; X_1 \cap X_i, \ldots, X_{i-1} \cap X_i, X_{i+1} \cap X_i, \ldots, X_n \cap X_i)$$

is an n-ad.

A map of n+1-ads $f: X \to Y$ is a family of maps $f(\alpha): X(\alpha) \to Y(\alpha)$ in the category such that $f(\alpha) = f|X(\alpha)$, where we write f itself for the $f(\alpha)$ when $\alpha = \{1, 2, \ldots, n\}$. Equivalently, f is a map $f: (X; X_1, \ldots, X_n) \to (Y; Y_1, \ldots, Y_n)$. f is allowable if $f^{-1}(Y_i) = X_i$, $i = 1, 2, \ldots, n$.

Definition 1.1. Let X be an n+1-ad. Given any α, we let $\partial X(\alpha) = \bigcup \{X(\beta) \mid \beta \subset \alpha \text{ and } \beta \neq \alpha\}$.

A Top (PL) manifold n+1-ad M is a Top (PL) n+1-ad such that $M(\alpha)$ is a Top (PL) manifold with boundary = $\partial M(\alpha)$. A component of $M(\alpha)$ is called a face of M.

For a smooth manifold n+1-ad we must first generalize the notion of smooth manifold with boundary to include "corners". (Cerf [ ]): We denote by $R^n_{(q)}$ any subset of R^n of the form $\{x \in R^n \mid x_i \geq 0$ for n-q distinct coordinates$\}$ and call it a model of type (n, q). Note that $R^n_{(n)} = R^n$ and $R^n_{(n-1)}$ is a half space H^n. $R^n_{(q)}$ for $q < n$ is homeomorphic, but not diffeomorphic, to H^n. A topological manifold (with boundary) M^n is said to have a smooth structure (with corners) if there is an atlas $\{h_\alpha\}$, $\alpha \in A$, $h_\alpha : U_\alpha \to X_\alpha$ a homeomorphism of an open set U_α in some $R^n_{(q)}$, $q = q(\alpha)$, onto the open set $X_\alpha \subset M$, such that $h^{-1}_{\alpha_2} h_{\alpha_1}$ is smooth, $\alpha_i \in A$. That is, if $X_{\alpha_1} \cap X_{\alpha_2} \neq \phi$, $h^{-1}_{\alpha_2} h_{\alpha_1}$ extends to a smooth map of an open neighborhood of $h^{-1}_{\alpha_1}(X_{\alpha_1} \cap X_{\alpha_2})$ in R^n into R^n. Two smooth structures are equivalent if there union is a smooth structure. A smooth manifold (with corners) is a topological manifold together with an equivalence class of smooth structures.

Now let V be a smooth manifold without corners or boundary of dimension d, and W_i, $i = 1, 2, \ldots, n$ a family of smooth submanifolds without corners or boundary, closed in V, dim $W_i = d-1$, and the W_i intersecting in general position (i.e., at a point of intersection the tangent subspaces are in general position). This is also referred to as a transverse family of submanifolds. The closure M of a component of $V - \bigcup W_i$ is a smooth manifold with corners, and $(M; M_1, \ldots, M_n)$, $M_i = M \cap W_i$, is an n+1-ad. A smooth manifold n+1-ad is the disjoint union of such n+1-ads. Note that for any α, $M(\alpha)$ is a smooth manifold k-ad for some k.

In general, we may define a q-face of a manifold with corners as the closure of a component of $\{x \in M^d \mid x$ is the origin of a chart of type (d, q)$\}$. If M is not a manifold n+1-ad, a face is not always a manifold with corners however.

Example 3. If M and N are smooth manifolds with boundary, $M \times N$ is a smooth manifold with corners. More generally, if M and N are smooth manifolds with corners, so is $M \times N$.

Example 4. If M is a smooth manifold with boundary,

$$(\Delta^{n-1} \times M, \partial_0 \Delta^{n-1} \times M, \ldots, \partial_{n-1} \Delta^{n-1} \times M, \Delta^{n-1} \times \partial M)$$

is an n+2-ad. More generally, if M is a smooth manifold k-ad,

$$(\Delta^{n-1} \times M, \partial_0 \Delta^{n-1} \times M, \ldots, \partial_{n-1} \Delta^{n-1} \times M, \Delta^{n-1} \times M_1, \ldots, \Delta^{n-1} \times M_{k-1})$$

is a smooth manifold n+k-ad.

Example 5. Let $C(\Delta^{n-1})$ be the infinite cone with vertex at the origin spanned by Δ^{n-1} in R^n. Then $C(\Delta^{n-1})$ is a smooth n+1-ad. In particular, $C(\Delta^0) = H^1$, the half line.

Remark. If $m \leq n$, we consider $R^m = \{x \in R^n \mid x_i = 0 \text{ for } i > m\}$. Then if for particular models we have $R^m_{(p)} \subset R^n_{(q)}$ we call $R^m_{(p)}$ a submodel of $R^n_{(q)}$.

Definition 1.2. Let M^m and N^n be manifolds (with corners). A map $f: M^m \to N^n$ is called an immersion if for each $x \in M$ there are charts $h: U \to M$, $h(U) \ni x$, U open in $R^m_{(p)}$ and $k: V \to N$, $k(V) \ni f(x)$, V open in $R^n_{(q)}$ such that $U = V \cap$ a submodel of type (m, p) and $f \circ h = k|U$. An immersion f is called an embedding if f is also a topological embedding.

Example 6. If $M \subset N^n$ is a subspace and for each $x \in M$ there is a chart $k: V \to N$, $k(V) \ni x$, V open in $R^n_{(q)}$ such that $k^{-1}(M) = V \cap$ a submodel of type (m, p), then M is a manifold with corners such that the inclusion is an embedding. We call M a submanifold.

If N is a manifold k-ad and $M \subset N$ is a submanifold such that with the induced k-ad structure, $M_i = M \cap N_i$, M is a manifold k-ad, then M is called a submanifold k-ad.

Remark. Definition 1.2 applies to the PL and Top categories as well. Since corners have no intrinsic meaning in these categories it is only necessary to consider $R^n_{(q)}$, $q = n, n-1$ for N. However, if say M is a manifold 3-ad with $\partial M = M_1 \cup M_2$ and we wish to consider immersions or embeddings $f: M \to W$ with $f^{-1}(\partial W) = M_1$, then we need a corner at $\partial M_1 = \partial M_2$, and so we may need $p = m, m-1, m-2$.

A PL or Top embedding of manifolds satisfying (1.2) is called a locally flat embedding.

Example 7. If M is a smooth manifold k-ad, then M may be embedded as a submanifold k-ad in $C(\Delta^{k-2}) \times R^s$, s sufficiently large.

Definition 1.3. Let $f: M \to N$ be a smooth map of smooth manifold n-ads. f is said to be transverse to the boundary if f and all the N_i are in general position; i.e., if $f(x) \in N(\alpha)$, $df_x(M_x)$ and $\bigcap (N_i)_{f(x)}$, $i \notin \alpha$, span $N_{f(x)}$.

If f is an immersion or an embedding, f is transverse to the boundary.

Lemma 1.4. Let V be a smooth manifold n-ad and X^j, $j = 1, \ldots, r$, smooth submanifold n-ads in general position and closed in V. Let W be a smooth manifold n-ad and $f: \bigcup X^j \to W$ an allowable map such that $f|X^j$ is smooth and transverse to ∂W, $j = 1, \ldots, r$. Then f extends to a smooth allowable map $F: U \to W$, U a neighborhood of $\bigcup X^j$ in M, transverse to ∂W.

Proof. Embed W in $C(\Delta^{n-2}) \times R^k$, transverse to the boundary. Then f may be considered as a map into $C(\Delta^{n-2}) \times R^k \subset R^{n-1} \times R^k$. Let $g = \pi_s \circ f$, $\pi_s: R^{n-1} \times R^k \to R$ the projection onto the sth factor. Then $g: \bigcup X^j \to R$ and $g|X^j$ is smooth. We claim it is sufficient to show that g extends to $G: V \to R$, G smooth and $G|V_i \equiv 0$ if $g|\bigcup_j X^j_i \equiv 0$.

First note that $C(\Delta^{n-2}) \times R^k = H \times H \times \ldots \times H \times R^k$ (n-terms). Hence if $g = \pi_s \circ f$, $s \leq n$, $g: \bigcup X^j \to H$, $g|\bigcup_j X^j_s \equiv 0$, g is transverse to 0, and $g^{-1}(0) = \bigcup_j X^j_s$. Hence if $G|V_s \equiv 0$, there is a neighborhood U of $\bigcup X^j$ such that G is transverse to 0, $G^{-1}(0) = U \cap V_s$ and $G: U \to H$.

Consequently the G's define a smooth map $\overline{F}: U \to C(\Delta^{n-2}) \times R^k$ extending f with $F^{-1}(C(\Delta^{n-2}) \times R^k)_i = U_i$ and transverse to the boundary. Since we may construct a tubular neighborhood T of W in $C(\Delta^{n-2}) \times R^k$ and a smooth retract $p: T \to W$ such that $p^{-1}(W_i) = T_i$ and p is transverse to the boundary, $F = p\overline{F}$ will satisfy the required properties for U sufficiently small that $\overline{F}(U) \subset T$.

The desired extension of g to G will follow from the Lemma below by adjoining V_s to the X^j's and taking $g \equiv 0$ in V_s.

Lemma 1.5 (R. Thom [32]): Let V be a smooth manifold without boundary and X^j, $j = 1, \ldots, r$, a family of smooth submanifolds without boundary, closed in V and in general position. Let $g: \bigcup_{j=1}^{r} X^j \to R$ be a map such that $g|X^j$ is smooth, $j = 1, \ldots, r$. Then g extends to a smooth map $G: V \to R$.

Proof. We proceed by induction on r. If $r = 1$, choose a smooth normal tube T of X_1 in W of radius a, and let $p: T \to X_1$ be a smooth retraction. Let $\lambda: [0, a] \to [1, 0]$ be a smooth non-decreasing function which is constant near the endpoints, $\lambda(0) = 1$, $\lambda(a) = 0$. Define $G(x) = \lambda(|x|)\lambda(p(x))$ for $x \in T$, $|x|$ = radius of x. Extend G outside of T by $G(x) = 0$.

Now suppose the result holds for $r-1$. Since the manifolds X_j, $j > 1$, cut X_1 transversally, we may choose a Riemannian metric on V and a normal tube T with respect to this metric so that $p(T \cap X_j) = X_1 \cap X_j$, $j > 1$. Let F be a smooth extension of $g|\bigcup X_j$, $j > 1$. Define h on X_1 by $h = g - F$. h is smooth and h is zero on $X_j \cap X_1$, $j > 1$. Extend h to a smooth function H on V as in the first paragraph. Then $H|X_j = 0$, $j > 1$. Hence $G = F + H$ satisfies the requirements.

Remark. The formalization of the notion of n-ads is due to Wall [18].

§ 2. Isotopies

a) Smooth Category

Let V and W be smooth manifolds. Two smooth embeddings f_0, f_1 $f_0, f_1: V \to W$ are called isotopic if one can deform f_0 to f_1 through embeddings. Actually there are several definitions of isotopy in the literature depending on the precise nature of the deformation: An isotopy may be viewed as a map $f: I \times V \to I \times W$ of the form $f(t, y) = (t, f_t(y))$, where $f_t: V \to W$ is a smooth embedding, $0 \leq t \leq 1$, with one of the following conditions imposed.

1. The partial derivatives of f_t depend continuously in t.
2. f is a smooth embedding of $I \times V$ in $I \times W$.
3. f is a smooth embedding and the partial derivatives of f at $t = 0$ $(t = 1)$ coincide with those of $id \times f_0$ $(id \times f_1)$.
4. f is a smooth embedding and $f = id \times f_0$ $(id \times f_1)$ for t near zero (t near one).

Obviously, $4 \Rightarrow 3 \Rightarrow 2 \Rightarrow 1$. On the other hand, if f_0 and f_1 are isotopic by an isotopy satisfying (1), we may clearly deform the isotopy so it is constant near zero and one; i.e., so it satisfies the second condition of (4). But the argument in Munkres [31], for deforming a "regular homotopy" to a "differentiable homotopy" says we may deform such an isotopy to one satisfying (4) so that the deformation of f_t for each t is a C^{∞}-δ homotopy. Thus the property that f_0 and f_1 are isotopic does not depend on the choice of the condition 1-4.

A p-isotopy is a map $f: \Delta^p \times V \to \Delta^p \times W$ of the form $f(x, y) = (x, f_x(y))$ where $f_x: V \to W$ is a smooth embedding, $x \in \Delta^p$, with an analogous further condition 1-4 imposed (see [10] and below for the higher analogues of (3) and (4)). The set of p-isotopies may be taken as the p-simplices of Δ-set , which turns out to be a Kan css set. The natural inclusions $4 \Rightarrow 3 \Rightarrow 2 \Rightarrow 1$ can be shown to be homotopy equivalences when V is compact. Condition (4) seems to be the most convenient for our purposes. On the other hand, (1) is the singular complex of the space of smooth embeddings $\underline{E}(V, W)$ with the (coarse) C^{∞} topology (see Munkres [31]). Consequently, we will impose the higher analogue of (4) and show the resulting css-set is a deformation retract of the singular complex $S\underline{E}(V, W)$.

In order to define this analogue we first note that a face $\sigma^q \subset \Delta^p$ has a neighborhood of the form $\sigma \times \Delta^{p-q}$. In fact, if τ^{p-q-1} is the opposite face to σ in Δ^p, identify $\partial_0 \Delta^{p-q}$ with τ and write a point in Δ^{p-q} in the form $tv_0 + (1-t)w$, $w \in \tau$, $0 \leq t \leq 1$. Then the linear isomorphism $sx + (1-s)w \to (x, 2sv_0 + (1-2s)w)$, $x \in \sigma$, $0 \leq s \leq 1/2$, identifies a neighborhood of σ with $\sigma \times \Delta^{p-q}$.

Now let $V = (V; V_1, \dots, V_{\ell-1})$ be a smooth manifold ℓ-ad and $W = (W; W_1, \dots, W_{k-1})$ a smooth manifold k-ad, $\ell \geq k$. Consider V as the k-ad $(V; V_1, \dots, V_{k-1})$ by ignoring V_i, $i \geq k$. (Then if $f: V \to W$ is an allowable k-ad map, $f(\text{Int } V_i) \subset \text{Int } W$, $i \geq k$.). Define a p-simplex of $E(V, W)$ as an embedding $f: \Delta^p \times V \to \Delta^p \times W$ which is an allowable map of p+k+1-ads such that:

(2.1) f commutes with projection on Δ^p, and

(2.2) for each $\sigma \subset \Delta^p$, f coincides on a neighborhood of $\sigma \times V$ with the product of $f|\sigma \times V$ and the identify map on Δ^{p-q}.

If $\lambda: \Delta^q \to \Delta^p$ is in css, define $\lambda^{\#}f: \Delta^q \times V \to \Delta^q \times W$ by

(2.3) $$\lambda^{\#}f(x, y) = (x, p_2 f(\lambda(x), y)).$$

This makes $E(V, W)$ a css-set.

If $f_0: V \to W$ is a vertex of $E(V, W)$ we will denote the pointed set by $E(V, W, f_0)$ when it is necessary to specify the base point. If M is a subset of V, we write $E(V, W \bmod M)$ or more precisely $E(V, W, f_0 \bmod M)$ for the sub-css set of simplicies which satisfy

(2.4) $$f | \Delta^p \times M = id \times f_0 | M.$$

We define analogously the css-group $A(V)$ for which the p-simplices are $p + \ell + 1$-ad automorphisms of $\Delta^p \times V$ satisfying (2.1) and (2.2); and similarly $A(V \bmod M)$ for the automorphisms fixed on $\Delta^p \times M$.

$A(V \bmod M)$ is Kan since any css group is Kan. We also have:

<u>Proposition 2.5.</u> For any subset M of V, $E(V, W \bmod M)$ is Kan and a deformation retract of $S\underline{E}(V, W \bmod M)$.

<u>Proof.</u> Let $f: \Delta^n \to S\underline{E}(V, W \bmod M)$, $f(\partial\Delta^n) \subset E(V, W \bmod M)$. Since $f | \partial\Delta^n \times V$ satisfies (2.2) we may extend it to $\overline{f}$ on $(\text{neigh } \partial\Delta^n) \times V$ so as to satisfy (2.2). That is, extend f as a product in a $\text{neigh}(\sigma \times V)$ in $\Delta^n \times V$, $\sigma \in \partial\Delta^n$, inductively with respect to $\dim \sigma$. Thus f is homotopic rel $\partial\Delta^n \times V \cup \Delta^n \times M$ to $g: \Delta^n \to S\underline{E}(V, W \bmod M)$ satisfying (2.2), and hence smooth in a $(\text{neigh } \partial\Delta^n) \times V$. By the Munkres argument ([ ], Ch I, §4), g is homotopic rel $(\text{neigh } \partial\Delta^n) \times V \cup \Delta^n \times M$ to a smooth n-simplex satisfying (2.2) and hence to a g in $E(V, W \bmod M)$.

Now let $f: \Lambda_{n+i} \to E(V, W \bmod M)$. Since $S\underline{E}(V, W \bmod M)$ is Kan, f extends to $g: \Delta^n \to S\underline{E}(V, W \bmod M)$. Since $g(\partial_i\Delta^n) \subset E(V, W \bmod M)$, we may apply the above argument to deform $g | \partial_i\Delta^n \times V$ rel $\partial(\partial_i\Delta^n) \times V$ and hence g rel $\partial\Delta^n \times V$ so that $g(\partial\Delta^n) \subset E(V, W \bmod M)$. Again by the above argument, g is homotopic rel $\partial\Delta^n \times V$ to g in $E(V, W \bmod M)$

Hence $E(V, W \bmod M)$ is Kan and a deformation retract of $S\underline{E}(V, W \bmod M)$.

Similarly we have:

<u>Proposition 2.6.</u> $A(W \bmod M)$ is a deformation retract of $S\underline{A}(W \bmod M)$, M any subset of W.

b. PL Category

Let V^m be a PL manifold ℓ-ad and W^n a PL manifold k-ad, $\ell \geq k$. It will be sufficient to consider only $\ell = k, k+1$, since corners have no intrinsic meaning in the PL category. Consider V and a k-ad by ignoring V_k if $\ell = k+1$.

Define $E(V, W)$ to be the css set whose p-simplices are allowable PL embeddings of p+k+1 ads $f: \Delta^p \times V \to \Delta^p \times W$ satisfying (2.1) and the local flatness condition:

(2.7) For any simplex Δ' linearly embedded in Δ^p, $f|\Delta' \times V$ is a locally flat embedding in $\Delta' \times W$.

Let $f_0: V \to W$ be a base point; i.e., an allowable locally flat embedding. If for convenience we identify V with $f_0(V)$, then (2.7) is implied by the local isotopy extension condition:

(2.8) For each $g \in V$ and $x \in \Delta^p$, there is a neighborhood B of g in W, a neighborhood A of x in Δ^p and an embedding $\bar{f}: A \times B \to \Delta^p \times W$ commuting with projection into Δ^p and extending f. That is, $\bar{f}|A \times B \cap V = f|A \times B \cap V$.

The isotopy extension theorem of section 4 shows that (2.7) and (2.8) are in fact equivalent. Further if $n \geq m+3$, the condition (2.7) is true for any allowable PL embedding [7].

We again define $\lambda^{\#}f$ by (2.3). $E(V, W, f_0)$ and $E(V, W, f_0 \text{ mod } M)$ are defined analogously to the smooth case. $A(V)$ is defined as the css-group whose p-simplices are allowable PL automorphisms of manifold $p+\ell+1$ ads, $\Delta^p \times V$, commuting with projection onto Δ^p. $A(V \text{ mod } M)$ as the sub css group whose p-simplices are fixed on $\Delta^p \times M$. That $E(V, W, f_0 \text{ mod } M)$ is a Kan css-set is immediate from the fact that Δ^p is PL homeomorphic to $I \times \Lambda_{p,i}$.

c. Top Category.

Let V^m and W^n be Top manifold k-ads. (What to do when V is a k+1-ad will be considered in section 4.) Let $f_0: V \to W$ be an allowable locally flat embedding.

Define $E(V, W, f_0)$ to be the css-set whose p-simplices are allowable embeddings of p+k+1 ads $f: \Delta^p \times V \to \Delta^p \times W$ satisfying (2.1) and the local isotopy extension condition (2.8). Define $E(V, W, f_0 \bmod M)$, $A(V)$, $A(V \bmod M)$ analogously to the PL-case. These are again Kan css sets ($\lambda^{\#}f$ given by 2.3).

Note that $A(V \bmod M)$ may be identified with the css-set of singular simplices of the topological group $A(V \bmod M)$ of allowable k-ad homeomorphisms of V, fixed on M, with the compact-open topology.

We will write $E^d(V, W)$, $E^{p\ell}(V, W)$, $E^t(V, W)$, etc., when it is necessary to distinguish the category.

d. PD isotopies

Let V be a smooth manifold k-ad and $f_0: K \to V$ a smooth triangulation; K is a PL manifold k-ad and for each α, $f_0 | K(\alpha)$ is a smooth triangulation of $V(\alpha)$. Define $A^{pd}(V, f_0)$ to be the css-set whose p-simplices are PD (piecewise differentiable) p+k+1-ad homeomorphisms $f: \Delta^p \times K \to \Delta^p \times V$ commuting with projection on Δ^p and with $\lambda^{\#}f$ defined by (2.3). When there is no confusion we write $A^{pd}(V)$ for $A^{pd}(V, f_0)$. Similarly, we write $A^{pd}(V \bmod M)$ for $A^{pd}(V, f_0 \bmod M)$, the css subset whose p-simplices satisfy $f | \Delta^p \times M = f_0 | \Delta^p \times M$.

<u>Lemma 2.9.</u> $A^{pd}(V, f_0 \bmod M)$ is Kan.

<u>Proof.</u> Δ^n is PL homeomorphic to $I \times \Lambda_{n,i}$. Consequently, given a PD homeomorphism $f: \Lambda_{n,i} \times K \to \Lambda_{n,i} \times V$ we may extend it to a PD homeomorphism $g: \Delta^n \times K \to \Delta^n \times V$ by setting $g = h \times id_V \circ id_I \times f \circ h^{-1} \times id_K$, where

$h: I \times \Lambda_{n,i} \to \Delta^n$ is the PL homeomorphism. If f commutes with projection on $\Lambda_{n,i}$, g commutes with projection on Δ^n. Hence $f: \Lambda_{n,i} \to A^{pd}(V \bmod M)$ extends to $g: \Delta^n \to A^{pd}(V \bmod M)$; and the set is Kan.

Now K is a PL manifold k-ad and we may identify $A^{p\ell}(K \bmod M)$ to a css-subset of $A^{pd}(V, f_0 \bmod M)$ unded the map $g \to f_0 \circ g$.

Proposition 2.10. If M is a PL-subspace of K, $A^{p\ell}(K, f_0 \bmod M)$ is a deformation retract of $A^{pd}(V, f_0 \bmod M)$.

Proof. Let $f: \Delta^p \times K \to \Delta^p \times V$ be in $A^{pd}(V, f_0 \bmod M)$ with $\mathrm{id} \times f_0^{-1} \circ f | \partial\Delta^p \times K$ PL and $f|\Delta^p \times M = \mathrm{id} \times f_0$. If V is compact, then for a sufficiently fine subdivision of (K, M), the linear approximation to f (see [8]) is a PL-homeomorphism and commutes with projection. Hence is PD isotopic to $f_0 \circ g$ by an isotopy commuting with projection and preserving the above conditions.

In the non-compact case this argument must be modified, see Putz [33] or [5].

§3. Pseudo isotopies

a. Smooth Category

Let W be a smooth manifold k-ad and V a smooth manifold ℓ-ad, $\ell \geq k$, considered as a k-ad. Consider p-simplices as in E(V, W) except that we do not assume $f: \Delta^p \times V \to \Delta^p \times W$ commutes with the projection on Δ^p. For a face $\sigma^q \subset \Delta^p$ we can define f by restriction since it is a map of p+k+1 ads. But for degeneracies, the formula (2.2) no longer makes sense.

Consider the ith degeneracy $s_i: \Delta^{p+1} \to \Delta^p$, and define a map $\chi_i: \Delta^p \times I \to \Delta^{p+1}$ by sending (x, t) into the point which divides the segment $s_i^{-1}(x)$ in the ratio t/1-t. Define $s_i^{\#} f: \Delta^{p+1} \times V \to \Delta^{p+1} \times W$ by

(3.1) $\quad s_i^{\#} f(\bar{x}, y) = (\chi_i(p_1 f(x, y), t), p_2 f(x, y)),\ \bar{x} \in \Delta^{p+1},\ y \in V$, where $(x, t) \in \chi_i^{-1}(\bar{x})$ (The result is independent of the choice.)

Lemma 3.2. $s_i^{\#} f$ is a smooth embedding.

Proof. It is sufficient to check for one i, say $s_p^{\#} f$. For this purpose consider $\Delta^{p+1} = (v_0, \ldots, v_p, v_{p+1})$ linearly embedded in R^{p+1} as follows: Embed $\partial_p \Delta^p$ linearly in R^{p-1}, and take $v_p = (0, \ldots, 0, 1, 0)$ and $v_{p+1} = (0, \ldots, 0, 1, 1)$. Writing $f_i(x, y)$ for the ith coordinate in R^{p+1} of $p_1 f(x, y)$, we have then $s_p^{\#} f(x_1, \ldots, x_{p+1}, y) = (f_1(x, y), \ldots, f_p(x, y), tf_p(x, y), p_2 f(x, y))$, $\overline{x} = (x_1, \ldots, x_{p+1}) \in \Delta^{p+1} \subset R^{p+1}$, $x = (x_1, \ldots, x_p, 0)$, $y \in V$, and where $x_{p+1} = tx_p$, $0 \leq t \leq 1$.

Note that $f_p(x, y) = 0$ if and only if $x_p = 0$. Since f is an embedding, we see immediately that $s_p^{\#} f$ is a topological embedding. Since for $x_p \neq 0$, $t = x_{p+1}/x_p$, $s_p^{\#} f$ is smooth and of maximum rank at any point $(\overline{x}, y)$ with $x_p \neq 0$. By condition (2.2), f coincides in a neighborhood of (x, y), such that $x_p = 0$, with the product of $f|_{x_p = 0}$ and the map sending $(1-s)x + sv_p$ to $(1-s)x' + sv_p$, $x' = p_1 f(x, y)$, $0 \leq s \leq 1$; in particular, $f_p(x, y) = x_p$ on this neighborhood and hence $tf_p(x, y) = x_{p+1}$. It follows that the $(p+1)^{st}$ coordinate of $s_p^{\#} f$ is smooth at $x_p = 0$, and that $s_p^{\#} f$ is smooth and of maximum rank at $x_p = 0$ as well as $x_p \neq 0$.

Remark. If f commutes with projection onto Δ^p, formula (3.1) reduces to (2.3).

Define $\widetilde{E}(V, W)$ as the css-set whose p-simplices are allowable maps of p+k+1 ads $f: \Delta^p \times V \to \Delta^p \times W$ such that f is a smooth embedding satisfying (2.2), with degeneracies defined by (3.1).

Define the css-sets $\widetilde{E}(V, W \bmod M)$ and the css-groups $\widetilde{A}(V)$ and $\widetilde{A}(V \bmod M)$ analogously.

b. PL (Top) Category

Let W be a PL (Top) manifold k-ad and V a PL (Top) manifold

ℓ-ad, $\ell = k, k+1$, considered as a k-ad. Let $\widetilde{E}(V, W)$ be the Δ-set whose p-simplices are allowable embeddings of p+k+1 ads $f: \Delta^p \times V \to \Delta^p \times W$ such that

(3.3) f is locally flat.

Remark. If $\dim W \geq \dim V+3$, then in the PL case (3.3) is always satisfied [7].

We define face operators by restriction. In the topological category we may define degeneracies by the formula (3.1). If f is PL, $s_i^{\#}f$ as given in (3.1) is not PL, in general. One defines similarly the Δ (css)-set $\widetilde{E}(V, W \bmod M)$ and the Δ (css)-groups $\widetilde{A}(V)$ and $\widetilde{A}(V \bmod M)$.

Note that $\widetilde{A}^d(V \bmod M)$, $\widetilde{A}^t(V \bmod M)$, being css-groups, are Kan. To prove that $\widetilde{A}^{p\ell}(V \bmod M)$ is Kan, we need only use again the fact that Δ^n is PL homeomorphic to $\Lambda_{n,i} \times I$. The same fact gives that $\widetilde{E}^{p\ell}(V, W \bmod M)$ and $\widetilde{E}^t(V, W \bmod M)$ are Kan. The proof that $\widetilde{E}^d(V, W \bmod M)$ is Kan will depend on showing that the restriction $\widetilde{A}^d(W \bmod M) \to \widetilde{E}^d(V, W \bmod M)$ is a Kan fibration ($\dim W \geq \dim V+3$). See #5.

c. PD pseudo isotopies.

Let V be a smooth manifold k-ad and $f_0: K \to V$ a smooth triangulation. Define $\widetilde{A}^{pd}(V, f_0)$ to be the Δ-set whose p-simplices are PL p+k+1-ad homeomorphisms $f: \Delta^p \times K \to \Delta^p \times V$, with face maps defined by restriction. Define $\widetilde{A}^{pd}(V, f_0 \bmod M)$ analogously. By the same argument as in (2.5), $\widetilde{A}^{pd}(V, f_0 \bmod M)$ is Kan, M a PL subspace of K. By the same argument as in (2.6), $A^{p\ell}(K, f_0 \bmod M)$ is a deformation retract of $A^{pd}(V, f_0 \bmod M)$.

§4 Isotopy Extension Theorem

a. Smooth Category.

Let W be a smooth manifold with corners and $f_0: V \to W$ an embedding. An embedding $f: V \to W$ is said to have the same incidence relations as f_0 if for any face F of W, $f(x) \in F$ if and only if $f_0(x) \in F$. Let W' be the union of faces of W, and let $V' = f_0^{-1}(W')$. Let $\underline{E}(V, W, f_0 \bmod V')$ be the space of embeddings $f: V \to W$ with the fine C^∞-topology such that f has the same incidence relations as f_0 and $f|V' = f_0|V'$. Further, let M be a submanifold of V, closed in V, and let $M' = M \cap V'$. The following theorem is due to Cerf [27] and Palais [34].

Theorem 4.1. The restriction map $\pi: \underline{E}(V, W, f_0 \bmod V') \to \underline{E}(M, W, f_0|M \bmod M')$ is a fibration.

Remarks. a) It follows from the covering homotopy property, that the image of π is the union of components of the range.

b) If N is any subset of M, $\pi: \underline{E}(V, W \bmod V' \cup N) \to \underline{E}(M, W \bmod M' \cup N)$ is a fibration.

c) If W is compact, $V = W$ and f_0 = identity, then $\underline{E}(V, W, f_0 \bmod V') = A(W \bmod W')$.

d) If V is compact, the coarse and fine C^∞ topology in $\underline{E}(V, W, f_0 \bmod V')$ coincide.

e) π in (4.1) induces a Kan fibration

$$S(\pi): S\underline{E}(V, W \bmod V') \to S\underline{E}(M, W \bmod M').$$

Let W be a smooth manifold k-ad, V a compact smooth manifold ℓ-ad, $\ell \geq k$, considered as a k-ad. Let $f_0: V \to W$ be an allowable embedding which we take to be an inclusion.

Theorem 4.2. The restriction map $p: A(W \bmod W') \to E(V, W \bmod V')$ is a Kan fibration.

Proof. Let $g: \Delta^p \times V \to \Delta^p \times W$ be a p-simplex of $E(V, W \bmod V')$ and suppose $\varphi: \Lambda \times W \to \Lambda \times W$ is a lift of $g|\Lambda \times V$. Since $A(W \bmod W')$ is Kan, there is a $\overline{\varphi}: \Delta^p \times W \to \Delta^p \times W$ in $A(W \bmod W')$ with $\overline{\varphi}|\Lambda \times W = \varphi$. Let $g_1 = \overline{\varphi}^{-1} \circ g$. Then g_1 is a p-simplex of $E(V, W \bmod V')$ such that $g_1|\Lambda \times V$ is the inclusion.

Suppose for the moment that W is compact; then by Theorem 4.1 and Remarks (b) and (e), $S(\pi)$: $S\underline{A}(W \bmod W') \to S\underline{E}(V, W \bmod W')$ is Kan. Hence there exists an $h: \Delta^p \times W \to \Delta^p \times W$ in $S\underline{A}(W \bmod W')$ such that $h| \Lambda \times W$ = identity $h| \Lambda \times W$ = identity and $h|\Delta^p \times V = g_1$. Actually we may assume $h|\text{neigh}\, \Lambda \times W$ = identity since $g_1 |$ neigh $\Lambda \times W$ = inclusion by (2.2).

Now g_1 extends to $\bar{g}_1: \Delta^p \times T \to \Delta^p \times W$ in $E(T, W \bmod T')$, where T is a tubular neighborhood of V in W, using a product metric on $\Delta^p \times W$. By the tubular neighborhood theorem [27], we may assume that $h = \bar{g}_1$ on $\Delta^p \times T$. Then by the Munkres argument (see 2.6) we may deform h to $h' \in A(W \bmod W')$ rel $\Delta^p \times (V \cup W')$. Then $f = \bar{\varphi} \circ h'$ is a lift of g such that $f| \Lambda \times W = \varphi$. Thus p is Kan -- provided W is compact.

But in any case, with g_1 as in paragraph one of the proof, the projection of $g_1(\Delta^p \times W)$ in W is contained in the interior of a compact submanifold W_0 of W. Hence we can apply the above argument with W_0 in place of W to obtain a lift h' of g which is the identity outside W_0 and in $\Lambda \times W \cup \Delta^p \times W'$. Then again $f = \bar{\varphi} \circ h'$ is a lift of g such that $f| \Lambda \times W = \varphi$.

Now assume M is a compact smooth manifold j-ad, $j \geq \ell$, and that M is embedded in V so that, forgetting M_i for $i \geq \ell$, the inclusion is an allowable ℓ-ad map. Now considering M as a k-ad, we have:

Theorem 4.3. The restriction map $p: E(V, W \bmod V') \to E(M, W \bmod M')$ is a Kan fibration.

Proof. Let $g: \Delta^p \to E(M, V \bmod M')$ and $\varphi: \Lambda \to E(V, W \bmod V')$ a lift of $g|\Lambda$. Let v_0 be the vertex of Λ and let $f_0 = \varphi(v_0)$. Since Λ is contractible it lifts to $\psi: \Lambda \to A(W \bmod W')$. Applying (4.2) with M in place of V, there exists $f: \Delta^p \to A(W \bmod W')$ lifting g and with $f| \Lambda = \psi$. Then the restriction $f_1: \Delta^p \to E(V, W \bmod V')$ of f is a lift of g with $f_1 | \Lambda = \varphi$. Hence p is Kan.

Remarks. a) If N is any subset of M then $p: E(V, W \bmod V' \cup N) \to E(M, W \bmod M' \cup N)$ is a Kan fibration. Similarly for (4.2) with N a subset of V.

b) The image of p is the union of components -- see below.

Definition 4.7. Let B be a Δ-set and let v be a zero simplex. The component of v is the Δ-subset consisting of all simplices of B whose vertices may be connected to v by a finite sequence of 1-simplices. That is, v' and v are connected if there exist 1-simplices $\sigma_1, \dots, \sigma_r$ such that v is a vertex of σ_1, v' of σ_r and for each i, σ_i and σ_{i+1} have a common vertex.
(If B is Kan, the components are in 1-1 correspondence with $\pi_0(B)$.)

Lemma 4.8. Let $p: E \to B$ be a Kan fibration. Then p(E) is theunion of components of B.

Proof. Suppose v_0 is a 0-simplex in B with $v_0 = p(\overline{v}_0)$. Then if σ is a 1-simplex with say $\partial_0\sigma = v_0$, $\partial_1\sigma = v_1$, it follows from the Kan condition on p, that there is a 1-simplex $\overline{\sigma}$ in E with $p(\overline{\sigma}) = \sigma$ and $\partial_0\overline{\sigma} = \overline{v}_0$; and hence $\partial_1\overline{\sigma} = \overline{v}_1$ where $p(\overline{v}_1) = v_1$. Consequently, all zero simplices in the component of a vertex v_0 of B are in the image of p.

Now let σ be any n-simplex in the component of v_0 and let v be a vertex of σ. Then there is a $\overline{v}$ with $p(\overline{v}) = v$. Further, any 1-simplex $\tau \subset \sigma$, containing v as a vertex, may be lifted to $\overline{\tau}$ with corresponding vertex $\overline{v}$. In particular, this gives a well-defined lift of each vertex of σ, since there is a unique τ connecting it to v.

Assume inductively that all k-1 simplices of σ and any k-simplex with vertex v may be lifted compatibly. Consider any othere k-simplex ρ. There exists a unique k+1 simplex $\delta \subset \sigma$ with face ρ and all other k-faces having v as a vertex. By the Kan condition on p, δ may be lifted to $\overline{\delta}$ to agree with the lift on all the k-faces containing v. In particular, this gives a compatible lift of ρ.

Thus all k-simplices and any k+1 simplex containing v may be lifted compatibly. Thus by induction, σ may be lifted, and p maps onto the component of v_0.

Corollary 4.9. Let $p: E \to B$ be a Kan fibration and suppose E is Kan. Then each component in the image of p is Kan.

b. PL Category

Let W be a manifold k-ad, V a compact manifold ℓ-ad, $\ell = k, k+1$, considered as a k-ad. Let $f_0: V \to W$ be an allowable locally flat embedding which we take to be an inclusion. The following is due to Hudson and Zeeman [7].

Theorem 4.10. The restriction map $p: A(W \bmod W') \to E(V, W \bmod V')$ is a Kan fibration.

As in (4.3), (4.10) implies for M a compact manifold ℓ or ℓ+1-ad in V

Theorem 4.11. The restriction map $p: E(V, W \bmod V') \to E(M, W \bmod M')$ is a Kan fibration.

Remarks. a) For any subset $N \subset V$, $p: E(V, W \bmod V' \cup N) \to E(M, W \bmod M' \cup N)$ is a Kan fibration. Similarly for 4.10 with $N \subset V$.

b) The image of p is the union of components

c) It is not necessary that V be compact (see proof of (4.2)).

d) Let W be a PL manifold k-ad and X a compact PL subspace k-ad. Let W' be as above and $X' = W' \cap X$. Assume $\dim W(\alpha) \geq \dim X(\alpha) + 3$ all $\alpha \subset \{1, \ldots, k-1\}$. Define $E(X, W \bmod X')$ to be the Kan css set whose p-simplices are maps of p+k+1-ads $f: \Delta^p \times X \to \Delta^p \times W$ such that f is an allowable PL embedding, $f | \Delta^p \times X'$ is the inclusion, and f commutes with projection onto Δ^p. Then Hudson has proved (4.10) with X in place of V. Also (4.11) for X in place of V follows by the same argument.

c. Germs of embeddings

Before stating the results in the Top case, we consider another way in which to make a manifold ℓ-ad M, $\ell > k$, into a k-ad. If for example, M is a 2-ad, i.e., a manifold with boundary, we can make M into a 1-ad by adding an open collar to ∂M. More generally, if $\partial M = M_1 \cup M_2$ we can form $M_\infty = M \cup M_2 \times [0,\infty)$ with M_2 identified to $M_2 \times 0$. Thus if M is an ℓ-ad and we let $M_1' = \bigcup M_i$, $i < k$ and $M_2' = \bigcup M_i$, $i \geq k$, then $M_\infty = M \cup M_2' \times [0,\infty)$ is a manifold k-ad. This works even in the smooth case since (see §1) if M is the closure of a component of $W - \bigcup V_i$, $i = 1,\ldots,\ell-1$, M_∞ may be considered as an open subset of the closure of a component of $W - \bigcup V_i$, $i = 1,\ldots,k-1$.

Let $M_a = M \cup M_2' \times [0,a) \subset M_\infty$. Let V be a manifold k-ad and $f_0: M_\infty \to V$ an allowable (locally flat) embedding. Then we consider the cas set of germs $E_\gamma(M,V)$ defined as follows: Let $f: \Delta^p \times M_a \to \Delta^p \times V$ be in $E(M_a, V, f_0|M_a)$. Define $f_1: \Delta^p \times M_{a_1} \to \Delta^p \times V$ and $f_2: \Delta^p \times M_{a_2} \to \Delta^p \times V$ to be equivalent if there exists an $a_3 \leq (a_1, a_2)$ such that f_1 and f_2 agree on $\Delta^p \times M_{a_3}$. A p-simplex of $E_\gamma(M,V)$ is an equivalence class of such embeddings. Since each $E(M_a, V, f_0|M_a)$ is Kan, it is clear that $E_\gamma(M,V)$ is Kan (see (b) in proof below).

Theorem 4.12. In the smooth or PL category, the restriction map $p: E_\gamma(M,V) \to E(M,V)$ is a homotopy equivalence if M is compact.

Proof. a) p is onto: Let $g: \Delta^p \times M \to \Delta^p \times V$ be in $E(M,V)$. By using an internal collar there is an embedding $h: M_a \to M$ such that $h|M: M \to M \subset V$ is isotopic to the inclusion. Thus $g\ \mathrm{id} \times h|M$ is the restriction of $g \circ \mathrm{id} \times h$ in $E(M_a, V)$. Since $p: E(M_a, V) \to E(M,V)$ is a Kan fibration, g is the restriction of a p-simplex in $E(M_a, V)$.

b) p is Kan: Given $g: \Delta^p \times M \to \Delta^p \times V$ and $f_j: \partial_j \Delta^p \times M_{a_j} \to \partial_j \Delta^p \times V$, $j \neq i$, such that $f_j | \partial_j \Delta^p \times M = g | \partial_j \Delta^p \times M$ and f_j and f_k agree on

$(\partial_j\Delta^p \cap \partial_k\Delta^p) \times M_b$, $b < (a_j, a_k)$; we can find an $a > 0$ so that all the f_j are defined on $\partial_j\Delta^p \times M_a$ and agree on the corners. The result follows since $p: E(M_a, V) \to E(M, V)$ is a Kan fibration.

c. p is a homotopy equivalence: A germ [f] in the fibre, $f: \Delta^p \times M_a \to \Delta^p \times V$ satisfies $f|\Delta^p \times M = id \times f_0$. If $f|\partial\Delta^p \times M_a = id \times f_0$, f represents a homotopy class of the fibre. Using an internal collar as in (a), f may be deformed to $f \circ id \times h = id \times (f_0 \circ h)$. This covers the deformation of $p \circ f = id \times f_0$ to $id \times (f_0 \circ h|M)$. It follows by the covering homotopy property that the homotopy class of [f] is trivial.

Remarks. a) The same result holds for $p: E_\gamma(M, V \bmod M'_\gamma) \to E(M, V \bmod M')$, where M'_γ = germ of M'_∞.

b) We will usually use E(M, V) to denote $E_\gamma(M, V)$ as well. In the Top category we do not know whether the isotopy extension theorem (4.10) holds in general when V is a compact manifold k+1-ad. Consequently, in this category E(V, W mod V') will always mean $E_\gamma(V, W \bmod V'_\gamma)$ except when dim V = dim W when a special argument may be used (see section d).

c) Part (a) of the proof shows that an allowable (locally flat) embedding $f_0: M \to V$ always extends to M_∞.

Also if V is a PL manifold k+1-ad and W is a smooth manifold k-ad of the same dimension and $f_0: V_\infty \to W$ is a PD embedding, $E^{pd}(V, W \bmod V')$ will mean $E^{pd}_\gamma(V, W \bmod V'_\gamma)$. If $g_0: K \to W$ is a smooth triangulation, and if $f_0 = g_0 h_0$, $h_0: V_\infty \to K$ a PL embedding, then as in § 2(d) we can consider $E^{p\ell}(V, K, h_0 \bmod V')$ as a css subset of $E^{pd}(V, W, f_0 \bmod V')$, and essentially the same argument shows:

Proposition 4.13. $E^{p\ell}(V, K, h_0 \bmod V')$ is a deformation retract of $E^{pd}(V, W_0, f_0 \bmod V')$.

d. Top Category

Let W be a manifold k-ad, V a compact manifold ℓ-ad, $\ell = k, k+1$. For $\ell = k$ let $f_0 : V \to W$ (for $\ell = k+1$ let $f_0 : V_\infty \to W$) be an allowable locally flat embedding which we take to be an inclusion. The following theorem is due to Edwards and Kirby [35] and Lees [36]:

Theorem 4.14. The restriction map $p: A(W \bmod W') \to E(V, W \bmod V')$ is a Kan fibration.

Remarks. a) The argument of [35] applies directly to the case $\ell = k$. However, since their argument is local it applies as well to $\ell = k+1$ and $E(V, W \bmod V') = E_\gamma(V, W \bmod V'_\gamma)$.

b) If $\dim V = \dim W$, V_k separates W. Now V_k is a k-ad and $p: A(W \bmod W') \to E(V_k, W \bmod V_k')$ is a Kan fibration. Consequently, given $g: \Delta^p \times V \to \Delta^p \times W$ and $f: \Lambda \times W \to \Lambda \times W$, $f|\Lambda \times V = g|\Lambda \times V$, there exists $\varphi: \Delta^p \times W \to \Delta^p \times W$ such that $\varphi|\Delta^p \times V_k = g|\Delta^p \times V_k$ and $\varphi|\Lambda \times W = f$. Define $\psi: \Delta^p \times W \to \Delta^p \times W$ by $\psi = \varphi$ outside $\Delta^p \times V$ and $\psi = g$ on $\Delta^p \times V$. This shows the isotopy extension theorem holds in this case and it follows that $E_\gamma(V, W \bmod V'_\gamma) \to E(V, W \bmod V')$ is a homotopy equivalence.

As in (4.3), Theorem 4.14 implies for M a compact manifold ℓ or $\ell+1$ ad in V

Theorem 4.15. The restriction map $p: E(V, W \bmod V') \to E(M, W \bmod M')$ is a Kan fibration.

§5. Pseudo Isotopy Extension Theorem

A 1-simplex $f: I \times V \to I \times W$ of $\widetilde{E}(V, W)$ is usually called a pseudo-isotopy or concordance. Since we will later also be concerned with the css-set $E(I \times V, I \times W)$ we will call the higher simplices of $\widetilde{E}(V, W)$ pseudo-isotopies and reserve the name concordance for elements of $E(I \times V, I \times W)$.

a. Smooth Category

Let W be a smooth manifold k-ad and V a compact manifold ℓ-ad, $\ell \geq k$, considered as a k-ad. Let $f_0: V \to W$ be an allowable embedding. Let W' and $V' = V \cap W'$ be as in §4. Also assume $\dim W \geq \dim V + 3$. The following theorem is due to Hudson [6].

Theorem 5. 1. Let f be a 1-simplex of $\widetilde{E}(V, W \bmod V')$, $f: I\times V \to I\times W$, f_0 = inclusion. Then there is a 1-simplex H of $A(I\times W \bmod I\times W')$, $H: I\times(I\times W) \to I\times(I\times W)$, such that H_0 = identity, $H_1 \; f = id_I \times f_0$ and $H|I\times 0\times W$ = identity.

Corollary 5. 2. Let f be a 1-simplex of $\widetilde{E}(V, W \bmod V')$ and h_0 a zero simplex of $\widetilde{A}(W \bmod W')$ such that $h_0|V = f_0$. Then there is a 1-simplex $h. I\times W \to I\times W$ of $\widetilde{A}(W \bmod W')$ such that $h|0\times W = h_0$ and $h|I\times V = f$.

Proof. Let $g = (id_I \times h_0^{-1})\circ f$. Then $g: I\times V \to I\times W$ and g_0 = inclusion. By (5. 1) there is an $H: I\times(I\times W) \to I\times(I\times W)$ such that H_0 = identity, $H_1\circ g = id_I \times g_0$ and $H_1 | 0\times W$ = identity. Then $H_1^{-1}\circ(id_I\times g_0) = g = (id_I\times h_0^{-1})\circ f$ and $h = (id_I\times h_0)\circ H_1^{-1}$ satisfies $h|0\times W = h_0$ and $h|I\times V = f$.

However, we also need h to satisfy (3. 1). But for a 1-simplex, (3. 1) simply means that h is the product map $id \times h_0$ near zero and $id\times h_1$ near one. Since f satisfies the corresponding property, one may deform h near zero and one, rel $I\times V$, to satisfy this condition by using the tubular neighborhood theorem.

Remark. (5. 2) may be proved directly using the s-cobordism theorem (see [10]).

The following generalization of (5. 2) was announced by Morlet.

Theorem 5. 3. The restriction map $p: \widetilde{A}(W \bmod W') \to \widetilde{E}(V, W \bmod V')$ is a Kan fibration.

Proof. Let $g: \Delta^n \times V \to \Delta^n \times V$ be an n-simplex of $\widetilde{E}(V, W \bmod V')$, and let $f_j: \Delta^{n-1} \times W \to \Delta^{n-1} \times W$, $j \neq i$, be compatible lifts of $g | \partial_j \Delta^n \times V$. Since $\widetilde{A}(W \bmod W')$ is Kan, there exists $f: \Delta^n \times W \to \Delta^n \times W$ such that $f | \partial_j \Delta^n \times W = f_j$. By replacing g by $f^{-1} \circ g$, we may assume $g | \Lambda \times V$ = inclusion and f_j = identity.

Now by (3.1), g is the inclusion over a neighborhood N of $\Lambda_{n,i}$ in Δ^n. Since Δ^n may be deformed by an isotopy fixed in a neighborhood $N_0 \subset N$ of Λ to an embedding in N, we see that lifting a pseudo-isotopy $g': I \times (\Delta^{n-1} \times V) \to I \times (\Delta^{n-1} \times W)$, where g' is the inclusion on a neighborhood of $0 \times \Delta^{n-1} \times V \cup I \times \partial\Delta^{n-1} \times V$ (see figure 1 below). But this follows by 5.2. Thus there exists $h \in \widetilde{A}(W \bmod W')$ such that $h | \partial_j \Delta^n = f_j$ and $p(h) = g$.

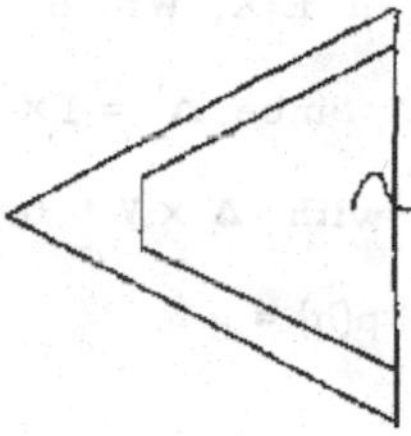

Remark. For any subset N of V, $p: \widetilde{A}(W \bmod W' \cup N) \to \widetilde{E}(V, W \bmod V' \cup N)$ is a Kan fibration.

By (4.9) we have:

Corollary 5.4. For any subset $M \subset V$, $\widetilde{E}(V, W \bmod M)$ is Kan.

b. PL Category.

Let W be a PL manifold k-ad and X a compact PL subspace k-ad. Let W' be as in §4 and $X' = X \cap W'$. Assume $\dim W(\alpha) \geq \dim X(\alpha) + 3$, all $\alpha \subset \{1, \ldots, k-1\}$. Define $\widetilde{E}(X, W \bmod X')$ to be the Kan Δ-set whose p-simplices are maps of p+k+1-ads $f: \Delta^p \times X \to \Delta^p \times W$ such that f is a proper PL embedding and $f | \Delta^p \times X'$ is the inclusion. The following theorem is due to Hudson [6].

Theorem 5.5. Let f be a 1-simplex of $\widetilde{E}(X, W \bmod X')$, $f: I \times X \to I \times W$, f_0 = inclusion. Then there is a 1-simplex H of $A(I \times W \bmod I \times W')$, $H: I \times (I \times W) \to I \times (I \times W)$, such that H_0 = identity, $H_1 \circ f = id_I \times f_0$ and $H|I \times 0 \times W$ = identity.

Corollalry 5.6. Let f be a 1-simplex of $\widetilde{E}(X, W \bmod X')$ and h_0 a zero simplex of $\widetilde{A}(W \bmod W')$ such that $h_0|X = f_0$. Then there is a 1-simplex $h: I \times W \to I \times W$ of $\widetilde{A}(W \bmod W')$ such that $h|0 \times W = h_0$ and $h|I \times X = f$.

The following generalization of (5.6) is due to Morlet [10].

Theorem 5.7. The restriction map $p: \widetilde{A}(W \bmod W') \to \widetilde{E}(X, W \bmod X')$ is a Kan fibration.

Proof. Let $g: \Delta^n \times X \to \Delta^n \times W$ be in $\widetilde{E}(X, W \bmod X')$ and let $f: \Lambda \times W \to \Lambda \times W$ be a lift of $g|\Lambda_{n,i} \times X$. Since $\Delta^n = I \times \Lambda$, we may apply (5.6) with W replaced by $\Lambda \times W$ and W' with $\Lambda \times W'$ to extend f to $h: \Delta^n \times W \to \Delta^n \times W$ in $A(W \bmod W')$ with $p(h) = g$.

c. Top Category

Let W be a topological manifold k-ad and V a proper compact submanifold k-ad embedded locally flatly, $\dim W \geq \dim V + 3$. Assume further that V is a topological handlebody ; i.e., each $V(\alpha)$ is built up from $\partial V(\alpha)$ by adding handles

The following theorem is due to E. Pederson [38]:

Theorem 5.8. Let f be a 1-simplex of $\widetilde{E}(V, W \bmod \partial V)$, $f: I \times V \to I \times W$, f_0 = inclusion. Assume $\dim W \geq 5$ and if $\dim W = 5$, that ∂W is a stable manifold. Then there is a 1-simplex H of $A(I \times W \bmod I \times \partial W)$, $H: I \times (I \times W) \to I \times (I \times W)$ such that H_0 = identity, $H_1 \circ f = id_I \times f_0$ and $H|I \times 0 \times W$ = identity.

Now let $W' = W(\alpha_1) \cup \ldots \cup W(\alpha_r)$, $\alpha_i \subset \{1, \ldots, k-1\}$ and $V' = W' \cap V$. Assume in addition to the hypothesis in the 1st paragraph that $W(\alpha)$ is a PL manifold-ad if $\dim W(\;\;) \leq 4$.

Corollary 5.9. Let f be a 1-simplex of $\widetilde{E}(V, W \bmod V')$, $f: I\times V \to I\times W$, f_0 = inclusion. Then there is a 1-simplex H of $A(I\times W \bmod I\times \partial W \cup 0\times W')$ such that H_0 = identity, $H_1 \circ f = id_I \times f_0$.

Proof. If $\dim W \le 4$, W is a PL manifold-ad by hypothesis, and since $\dim V \le 1$, V is a PL manifold. Now we may ε-tame V in W (see [37]); in particular there is an ambient isotopy k of W sending V to a PL embedding. Taking the product ambient isotopy of $I\times W$, we may assume V is a PL submanifold of W. By ε-taming modulo the boundary, we may assume f is a PL embedding and apply (5.5). Thus we obtain H for the case $\dim W \le 4$.

If $\dim W \ge 5$, we may inductively apply (5.8) to $(W(\alpha), \partial W(\alpha))$ and then extend the resulting isotopy of $I\times W(\alpha)$ to $I\times W$ (by inductively using the collar of $\partial W(\beta)$ in $W(\beta)$). Thus we get a composition H of a finite sequence of isotopies of $I\times W$ such that H satisfies the desired properties.

Corollary 5.10. With the hypothesis of 5.9, let f be a 1-simplex of $\widetilde{E}(V, W \bmod V')$ and h_0 a vertex of $\widetilde{A}(W \bmod W')$ such that $h_0|V = f_0$. Then there is a 1-simplex $h: I\times W \to I\times W$ of $\widetilde{A}(W \bmod W')$ such that $h|0\times W = h_0$ and $h|I\times V = f$.

By the same argument as (5.7) we get with the hypothesis of (5.9)

Theorem 5.11. The restriction map $p: \widetilde{A}(W \bmod W') \to \widetilde{E}(V, W \bmod V')$ is a Kan fibration.

Remark. By formally the same argument as we derived 4.12 from 4.11 we may obtain versions of 4.12 with $\widetilde{E}$ in place of E, for each category.

Definition. An embedding $f: X \to W$ of a PL complex in a topological manifold is called locally tame if for each $x \in X$ there is a neighborhood U of $f(x)$ in W and a PL manifold structure on U so that $f^{-1}(U) \to U$ is a PL embedding.

The following theorem is also due to E. Pederson [38].

Theorem 5.12. Let W^n be a topological manifold, $n \geq 5$, and if $n = 5$ assume ∂W is stable. Let X^p be a finite PL complex with a subcomplex X' (possibly empty) with $n-p \geq 3$. Assume that $f: I \times X \to I \times W$ is a map of 4-ads and a proper locally tame embedding; i.e., $f^{-1}(I \times W, 0 \times W, 1 \times W) = (I \times X, 0 \times X, 1 \times X)$ and $f^{-1}(I \times \partial W) = I \times X'$. Then there is an ambient isotopy H of $I \times W$ fixing $I \times \partial W \times 0 \times W$ such that $H_1 \circ f = id_I \times f_0$.

§6. Description of Homotopy Groups

An element of $\pi_n(\tilde{E}^d(V, W, f_0 \bmod M)$ is represented by an n-simplex $f: \Delta^n \times V \to \Delta^n \times W$ in $\tilde{E}^d(V, W, f_0 \bmod M)$ such that $f|\partial\Delta^n \times V = id \times f_0$. Two such f, f' represent the same class if and only if there exists a $g: \Delta^{n+1} \times V \to \Delta^{n+1} \times W$ in $\tilde{E}^d(V, W, f_0 \bmod M)$ such that $g|\partial_n \Delta^{n+1} \times V = f$, $g|\partial_{n+1}\Delta^{n+1} \times V = f'$ and $g|\partial_i \Delta^{n+1} \times V = id \times f_0$ for $0 \leq i < n$. This last we claim is the same as saying there exists a p+k+3-ad embedding $h: I \times \Delta^n \times V \to I \times \Delta^n \times W$ such that $h_t = f$ near zero and f' near one, $h|I \times \Delta^n \times M = id \times f_0$ and for some neighborhood U of $\partial\Delta^n$ in Δ^n, $h|I \times U \times V = id \times f_0$.

To see this we identify $I \times \Delta^n$ with the subspace of Δ^{n+1} defined as follows: Write $\Delta^{n+1} = \Delta^1 * \Delta^{n-1}$, $\overline{x} = (1-t)y + tz$, $0 \leq t \leq 1$, $\overline{x} \in \Delta^{n+1}$, $y \in \Delta^1$, $z \in \Delta^{n-1}$; and $\Delta^n = \Delta^0 * \Delta^{n-1}$, $x = (1-s)v + sz$, $0 \leq s \leq 1$, $v = \Delta^0$, $z \in \Delta^{n-1}$, $x \in \Delta^n$. Define $\phi_\varepsilon : I \times \Delta^n \to \Delta^{n+1}$ by $\phi_\varepsilon(y, x) = (1-\lambda_\varepsilon(s))y + \lambda_\varepsilon(s)z$, where $\lambda_\varepsilon : [0,1] \to [0, 1-\varepsilon]$ is a diffeomorphism such that $\lambda =$ identity on $[0, 1-3\varepsilon]$ and $\frac{d\lambda}{ds} = 1$ on $[1-\varepsilon, 1]$. Further, adjust the product structure on $I \times \Delta^n$ near $0 \times \Delta^n$ and $1 \times \Delta^n$ so that $\phi_\varepsilon|(\text{neigh } \partial I) \times \Delta^n$ is the product map, for $x \in \Delta^n$ with $0 \leq s \leq 1-2\varepsilon$, from this structure to the product structure near $\partial_n\Delta^{n+1}$ and $\partial_{n+1}\Delta^{n+1}$ in Δ^{n+1} used in (2.2).

Taking Δ^{n-1} to be $\partial_n\Delta^{n+1} \cap \partial_{n+1}\Delta^{n+1}$ and observing that $g_{\bar{x}} = f_0$ for $\bar{x}$ in a neighborhood of Δ^{n-1} in Δ^n by (2.2), we see that for ε sufficiently small that g defines a map h as above by restriction. Conversely, given h as above define $g = h$ over $\phi_\varepsilon(I \times \Delta^n)$ and f_0 over the rest of Δ^{n+1}. Then for ε sufficiently small g satisfies the above properties, establishing our claim.

Similarly, an element of $\pi_n(\tilde{E}^d(V, W, f_0 \bmod M), \tilde{E}^d(V, W, f_0 \bmod M))$ is represented by $f: \Delta^n \times V \to \Delta^n \times W$ in $\tilde{E}^d(V, W, f_0 \bmod M)$ such that f commutes with projection over $\partial_0\Delta^n$ and $f|\partial_i\Delta^n \times V = id \times f_0$, $1 \le i \le n$. Two such f, f' represent the same class if and only if there exists $g: \Delta^{n+1} \times V \to \Delta^{n+1} \times W$ in $\tilde{E}^d(V, W, f_0 \bmod M)$ such that g commutes with projection over $\partial_0\Delta^{n+1}$, $g|\partial_n\Delta^{n+1} \times V = f$, $g|\partial_{n+1}\Delta^{n+1} \times V = f'$ and $g|\partial_i\Delta^{n+1} \times V = id \times f_0$ for $1 \le i < n$. Again note that $g_{\bar{x}} = f_0$ for $\bar{x}$ in a neighborhood of Δ^{n-1} in Δ^n. Hence the same argument shows that f and f' represent the same class if and only if there exists $h: I \times \Delta^n \times V \to I \times \Delta^n \times W$ such that $h_t = f$ near zero and f' near one, h commutes with projection over $I \times \partial_0\Delta^n$ and for any $\sigma^q \subset \Delta^n$, $h|I \times (\text{neigh } \sigma) \times V =$ product of $h|I \times \sigma \times V$ and $id_{\Delta^{n-q}}$.

Remark. The above also shows that two representations f and f' of $\pi_{n-1}(E(V, W \bmod M))$ are equivalent if and only if there exists $h: I \times \Delta^{n-1} \times V \to I \times \Delta^n \times W$ such that h commutes with projection over $I \times \Delta^{n-1}$, $h_t = f$ near zero and f' near one, and for $\sigma^q \subset \Delta^{n-1}$, $h|I \times \text{neigh } \sigma \times V =$ product of $h|I \times \sigma \times V$ and $id_{\Delta^{n-q-1}}$.

In fact, if V is compact and M is a submanifold, closed in V, we can go further: Consider smooth n+k+1-ad embeddings $f: \Delta^n \times V \to \Delta^n \times W$, and with $f|\partial\Delta^n \times V \cup \Delta^n \times M = id \times f_0$. Define two such f, f' to be equivalent if there exists an $h: I \times \Delta^n \times V \to I \times \Delta^n \times W$, a smooth n+k+3-ad embedding such that $h_0 = f$, $h_1 = f'$, $h|I \times \Delta^n \times M \cup I \times \partial\Delta^n \times V = id \times f_0$. We claim the equivalence classes are in one-to-one correspondence with $\pi_n(\tilde{E}^d(V, W, f_0 \bmod M))$.

First note that by the tubular neighborhood theorem (§7) f as above can be deformed rel $\partial\Delta^n \times V \cup \Delta^n \times M$ to $\tilde{f} \in \tilde{E}^d(V, W, f_0 \bmod M)$; i.e., there is a k satisfying the properties of h above, except that $k_0 = \tilde{f}$ and $k_1 = f$. Similarly, there is a k' with $k_1' = \tilde{f}' \in \tilde{E}^d(V, W, f_0 \bmod M)$ and $k_0' = f'$. Thus we have an $\tilde{h}$ with $\tilde{h}_0 = \tilde{f}$ and $\tilde{h}_1 = \tilde{f}'$. Again by the tubular neighborhood theorem, we can assume that $\tilde{h}$ satisfies the properties of h in the first paragraph of this section. Our claim follows.

Similarly, for the relative groups we can ignore condition (2.2) both for the representatives and for the concordance between them. In fact, let $E_0(V, W, f_0 \bmod M)$ be the css-set of embeddings satisfying (2.1) but not (2.2). It follows from the proposition below and (2.5) that $E(V, W, f_0 \bmod M)$ is a deformation retract of $E_0(V, W, f_0 \bmod M)$. Hence if $f: \Delta^n \times V \to \Delta^n \times W$ is an n+k+1-ad embedding such that $f|\Delta^n \times M = \mathrm{id} \times f_0$ and $f|\partial_i\Delta^n \times V = \mathrm{id} \times f_0$, $1 \leq i < n$, and f commutes with projection over $\partial_0\Delta^n$, i.e., $f|\partial_0\Delta^n \times V$ is in $E_0(V, W, f_0 \bmod M)$, we can deform $f|\partial_0\Delta^n \times V$ into $E(V, W, f_0 \bmod M)$ and by the isotopy extension theorem we may deform f to f' so that $f'|\partial_0\Delta^n \times V$ is in $E(V, W, f_0 \bmod M)$. Then by the tubular neighborhood theorem we may deform f' rel $\partial\Delta^n \times V \cup \Delta^n \times M$ so as to satisfy (2.2).

The argument of the proposition and the tubular neighborhood theorem can also be used to deform the resulting concordance commuting with projection over $I \times \partial_0 \Delta^n$ obtained as above to one satisfying the required local product conditions along the boundary.

Proposition 6.1. If V is compact and M is a submanifold, closed in V, $E_0(V, W \bmod M)$ is Kan and a deformation retract of $S\underline{E}(V, W \bmod M)$.

Proof. Let $f: \Delta^n \to S\underline{E}(V, W \bmod M)$, $f|\partial_i\Delta^n \times V$ in $E_0(V, W \bmod M)$. By (1.5), $p_2 f: \bigcup \partial_i\Delta^n \times V \to \Delta^n \times W \to W$ extends to a smooth map on $U \times V$ agreeing with $f_0\, p_2$ on $U \times M$, U a neighborhood of $\partial\Delta^n$ in Δ^n. Hence for U sufficiently small $(p_1, p_2 f): U \times V \to U \times W$ is a smooth embedding commuting with projection. We may deform f rel $\partial\Delta^n \times V \cup \Delta^n \times M$ so as to agree with $(p_1, p_2 f)$ over a neighborhood $\partial\Delta^n$. Now by the Munkres argument f may be deformed rel $U_1 \times V \cup \Delta^n \times M$ to g in $E_0(V, W \bmod M)$.

That $E_0(V, W \bmod M)$ is Kan follows as in (2.5).

§7. Tubular Neighborhood Theorem (smooth category) [27].

Let V be a smooth manifold (with corners) and M a submanifold. If M has an interior point in a face Σ of V and M is connected, then $M \subset \Sigma$. Consequently, there is a face of smallest dimension containing a connected M, called the support of M.

For any face Σ of V, let $\partial\Sigma = \{ y \in \Sigma \mid y \in \text{face of lower dimension} \}$.

Definition 7.1. The relative boundary of M in V, ∂M_V, is defined as follows: First if M is connected and Σ is the support of M, $\partial M_V = \{ x \in \partial M \mid x \notin \partial\Sigma \}$. For a general M, ∂M_V is the union of the relative boundaries of the components.

Let W be a smooth manifold (with corners) and $f: V \to W$ an embedding. Assume M is a submanifold of V, closed in V and of positive codimension in V. Define $J^r_M E(V, W, f)$ $(J^r_M E(V, W, f \bmod M))$ as the quotient space of $\underline{E}(V, W, f)$ $(\underline{E}(V, W, f \bmod M))$ defined by the relation: f' and f" are equivalent if f' and f" have contact of order r at each point of M. This implies, in particular, that f' and f" coincide in M.

Theorem 7.2. $\underline{E}(V, W, f) \to J^r_M E(V, W, f)$ and $\underline{E}(V, W, f \bmod M) \to J^r_M E(V, W, f \bmod M)$ as fibrations.

Addendum to Theorem 7.1. Let L be a normal tube in V over a closed subset of M. Then $E(V, W, f \bmod M \cup L) \to J^r_M E(V, W, f \bmod M \cup L)$ is a fibration.

Remark. Let $E_{J^r_M}(V, W, f)$ be the subspace of $\underline{E}(V, W, f)$ of embeddings $f': V \to W$ having contact of order r with f along M. Similarly, for $E_{J^r_M}(V, W, f \bmod M)$, etc. (Note that for $r = 1$, $E_{J^1_M}(V, W, f \bmod M)$ is the subspace of embeddings $f': V \to W$ such that $df = df'$.) Then $E_{J^r_M}(V, W, f) = E_{J^r_M}(V, W, f \bmod M)$ and $E_{J^r_M}(V, W, f \bmod M \cup L)$ are the fibres of the three fibrations above.

Let M be compact an d let $R(TV|M, TW, df \bmod TM)$ be the space of embeddings (with the coarse C^∞ topology) of $TV|M$ into TW which are bundle monomorphisms over f which agree with df on $TM \subset TV|M$. (This is the same as a space of smooth sections of an associated bundle to $TV|M$ with fibre a Stiefel manifold.)

Proposition 7.3. Let M be compact and let N be a normal tube of M in V. $d: \underline{E}(N, W, f \bmod M) \to R(TV|M, TW, df \bmod TM)$ induces a homeomorphism $\bar{d}$ of $J^1_M E(N, W, f \bmod M)$ with $R(TV|M, TW, df \bmod TM)$.

Proof. d is continuous. Now we may identify N with the set of vectors of length $\leq \varepsilon$ in $TV|M$ which are perpendicular to TM. Then given a bundle monomorphism $\varphi: TV|M \to TW$ such that $\varphi|TM = df|TM$, $\exp \circ \varphi|N$ embeds N in W. Hence (for ε sufficiently small), d has local cross-section. Since $\bar{d}$ is injective, it is a homeomorphism.

Theorem 7.4. $\pi_i(E_{J^1_M}(N, W, f \bmod M)) = 0$ all $i \geq 0$. Hence d in Proposition 2 induces an isomorphism on π_i, $i \geq 0$.

Corollary 7.5. Let $p: \underline{E}(V, W, f \bmod M) \to \underline{E}(N, W, f \bmod M)$ be the restriction fibration (4.3). Since $dp = d: \underline{E}(V, W, f \bmod M) \to R(TV|M, TW, df \bmod TM)$, the latter is a fibration since the fibre of p over $f'|N$ is $\underline{E}(V, W, f' \bmod N)$, we have the exact sequence:

$$\to \pi_i(\underline{E}(V, W, f' \bmod N), f') \to \pi_i(\underline{E}(V, W, f \bmod M), f')$$

$$\to \pi_i(R(TV|M, TW, df \bmod TM), f') \to \pi_{i-1}(\underline{E}(V, W, f' \bmod N, f') \to$$

$$\cdots \to \pi_0(\underline{E}(V, W, f \bmod M), f') \to \pi_0(R(TV|M, TW, df \bmod TM), df') \ .$$

<u>Corollary 7.5'.</u> Taking $V = W$ and replacing $\underline{E}(V, W, f \bmod M)$ by $\underline{A}(W \bmod M)$ we get the exact sequence:

$$\to \pi_i(\underline{A}(W \bmod N)) \to \pi_i(\underline{A}(W \bmod M)) \to \pi_i(R(TW|M, TW, \bmod TM))$$

$$\to \pi_{i-1}(\underline{A}(W \bmod N)) \to \cdots \to \pi_0(R(TW|M, TW, \bmod TM)) \ .$$

<u>Corollary 7.6.</u> The inclusion $\underline{E}(V, W, f \bmod N) \to E_{J^1_M}(V, W, f)$ induces an isomorphism on π_i, $i \geq 0$, for any base point $f' \in \underline{E}(V, W, f \bmod N)$.

<u>Corollary 7.6'.</u> The inclusion $\underline{A}(W \bmod N) \to A_{J^1_M}(W)$ induces an isomorphism on π_i, all $i \geq 0$.

<u>Proof.</u> This follows from the commutative diagram:

$$\begin{array}{ccccc}
 & \nearrow & E_{J^1_M}(V, W, f) & & \\
 & & \downarrow & & \\
\underline{E}(V, W, f \bmod N) & \to & \underline{E}(V, W, f \bmod M) & \xrightarrow{p} & \underline{E}(N, W, f \bmod M) \\
 & & \downarrow q & & \downarrow q \ \simeq \\
 & & J^1_M E(V, W, f \bmod M) & \xrightarrow{\bar{p}} & J^1_M E(N, W, f \bmod M)
\end{array}$$

and the fact that from

$$\begin{array}{ccc}
\underline{E}(V, W, f \bmod M) & \xrightarrow{p} & p(\underline{E}(N, W, f \bmod M) \\
\downarrow q & \swarrow & \downarrow q \ \simeq \\
J^1_M E(V, W, f \bmod M) & \overset{\bar{p}}{=} & qp(\underline{E}(N, W, f \bmod M) \subset J^1_M E(N, W, f \bmod M) \ ,
\end{array}$$

Similarly for $\underline{A}(W \bmod N)$.

Also we get the corresponding results for $\underline{E}(V, W, f \bmod M \cup L)$ and $\underline{A}(W, \bmod M \cup L)$. For example:

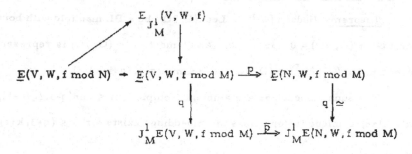

Corollary 7.6". The inclusion $\underline{A}(W \bmod N \cup L) \to A_{J^1_M}(W \bmod M \cup L)$ induces an isomorphism on π_i, all $i \geq 0$.

Of course, we can replace all of the above spaces by their singular complexes. Then using (2.5) we can also replace $SE(V, W, f \bmod M)$ by $E(V, W, f \bmod M)$, etc. Thus we get for example:

Corollary 7.$\overline{5}$. $\to \pi_i(E(V, W, f \bmod N) \to \pi_i(E(V, W, f \bmod M) \to \pi_i(SR(TV|M, TW, df \bmod TM)) \to \pi_{i-1}(E(V, W, f \bmod N)) \to \cdots \to \pi_0(SR(TV|M, TW, df \bmod TM))$.

Corollary 7.$\overline{6}$. $E(V, W, f \bmod N)$ is a deformation retract of $E_{J^1_M}(V, W, f)$.

§8. Hudson's Embedding Theorem

Theorem (Hudson [7]). Let $(Q, \partial Q)$ be a PL manifold with boundary such that $\pi_i(Q, \partial Q) = 0$ for $i \leq k$. An element of $\pi_r(Q, \partial Q)$ is represented by an embedding $f: (D^r, \partial D^r) \to (Q, \partial Q)$ if $r \leq \min(q-3, (q+k-1)/2)$.

Two such embeddings are ambient isotopic if $r \leq \min(q-3, (q+k-1)/2)$. In particular, a unique isotopy class of embeddings exists for $r \leq (q-3, k+1)$.

Corollary. Let $(Q, \partial Q)$ be a smooth manifold with boundary such that $\pi_i(Q, \partial Q) = 0$ for $i \leq k$. Then

a) An element of $\pi_r(Q, \partial Q)$ is represented by a smooth embedding $f: (D^r, \partial D^r) \to (Q, \partial Q)$ if $r \leq \min(q-3, (q+k-1)/2)$.

b) If $g: (D^r, \partial D^r) \to (Q, \partial Q)$ is an embedding homotopic to f and $\varepsilon > 0$, then f is ambient isotopic to a smooth embedding ε-close to g, if $r \leq \min(q-3, (q+k-2)/2)$.

c) f is ambient isotopic to g if $r \leq \min(q-3, 2(q-2)/3, (q+k-2)/2)$.

In particular, if $r \leq k+1$, the conditions for the above become

a) $r \leq q-3$, b) $r \leq q-3$, c) $r \leq \min(q-3, 2(q-2)/3)$

Proof. Starting with the PL embeddings and isotopies given by the theorem and the fact that the normal block bundle of a disc is trivial, a) and b) follow by the Cairns-Hirsch theorem [5]. c) follows from b) by Haefliger [39].

Appendix 2 -- The Topological Category

E. Pedersen

In this appendix we will extend a number of results to the topological category. First we consider Morlet's lemma of disjunction: We succeed to the following extent.

Theorem. Let V^n be a topological manifold, $n \geq 5$, and if $n = 5$, assume ∂V is stable. Let $g: (D^p, \partial D^p) \to (V, \partial V)$ and $h: (D^q, \partial D^q) \to (V, \partial V)$ be disjoint embeddings (locally flat) of discs with $n-p \geq 3$ and $n-q \geq 3$. Then

$$\pi_i(C(D^p, V; g), C(D^p, V-h(D^q); g) = 0$$

for $i \leq 2n-p-q-5$.

We also generalize the addendum.

Addendum. If V^n is simply connected, then the result holds for $i \leq 2n-p-q-4$.

Proof of Theorem. We note that with the added assumption on V^n, everything in the PL (Diff) proof holds true up to Proposition A', replacing Hudson's theorem by [38], so the main problem is transversality. For the solution of the transversality problem we note that Proposition A' is only needed for the case where $W_i = D^{p_i}$, so replacing Hudson's theorem by [38], the proof goes directly through (also using embedding results due to J. Lees) once the following transversality considerations have been made:

§1. Transversality in PL and topological manifolds.

We consider the problem of moving submanifolds to transversally intersecting submanifolds in PL and topological categories. When there are only two submanifolds, this presents no problem, using PL blocktransversality, but when there are more than two submanifolds one encounters particular problems in the topological category, namely whether or not blocktransversality is preserved under change of PL-structure.

The main motivation for these considerations is the following situation which arises in "Morlet's lemma of disjunction". Given embeddings

$$\varphi_i : (D^{p_i}, \partial D^{p_i}) \times (I, 0, 1) \to (V, \partial V) \times (I, 0, 1),$$

denote $\varphi_i | D^{p_i} \times 0$ by g_i and assume

$$\varphi_i | \partial D^{p_i} \times I = (g_i | \partial D^{p_i}) \times 1_I$$

and $n - p_i \geq 3$. Find small ambient isotopies h_i^t satisfying that the restriction to $\partial V \times I$ is a productisotopy, and such that φ_i are moved to "transversal" embeddings.

We present a solution here to the topological problem which might also have some interest in the PL case.

First we consider the situation in the PL category: If U is a PL manifold and M_i are PL submanifolds, they are said to be locally transversally intersecting if every point in U has a neighborhood PL homeomorphic to R^n by a homeomorphism that sends $M_i \cap U$ to subvectorspaces in general position, i.e., sends $M_{i_1}^{n_1} \cap M_{i_2}^{n_2} \cap \ldots \cap M_{i_s}^{n_s}$ to a subvectorspace of R^n of dimension $n_1 + n_2 + \ldots + n_s - (s-1)n$. In the case of a manifold with boundary, we replace R^n by H^n, halfspace, in the above definition and consider ∂U to be one of the submanifolds.

We note that this definition of transversality is a local one and different from the concept of blocktransversality by an example due to Hudson [40] (see also [41]).

Let M_i, $i \in J$, J finite, be locally transversally intersecting PL submanifolds in a PL manifold P, and denote $\bigcap_{s \in J'} M_s$, $J' \subseteq J$, by $M_{J'}$ and $M_{\{\emptyset\}} = P$. Whenever $J'' \subset J'$ we have $M_{J'} \subset M_{J''}$ and a neighborhood of $M_{J'}$

in $M_{J''}$ has the structure of a blockbundle $\nu_{J',J''}$ the normal blockbundle, see [25]. The blockbundle is described by giving a cellstructure to $M_{J'}$ and to each cell assign a block over the cell, the procedure of [25] being: triangulate so that $M_{J'} \subset M_{J''}$ is the inclusion of a full subcomplex, give $M_{J'}$ the cellstructure of dual cells, and take as block over the dual cell $C_\sigma^{M_{J'}}$ of σ in $M_{J'}$ the dual cell $C_\sigma^{M_{J''}}$ of σ in $M_{J''}$. We use the notation $\nu_{J'}$ for $\nu_{J',\{\emptyset\}}$, the normal blockbundle of $M_{J'}$ in P.

<u>Definition 1.</u> Notation as above: The normal blockbundles $\nu_{J',J''}$ of $M_{J'}$ in $M_{J''}$, $J'' \subset J'$ are said to be compatible if for every $J''' \subset J'' \subset J'$ and every cell C in $M_{J'}$ we have

1. $\nu_{J',J''}(C)$ is a cell in $M_{J''}$
2. $(\nu_{J'',J'''}(\nu_{J',J''}(C)), C) = (\nu_{J',J'''}(C), C)$

The property of being compatible can clearly be preserved under subdivision: In other words, if you subdivide the cell structure of $M_{J'}$ you can subdivide all $M_{J''}$ in such a way that conditions 1 and 2 above are preserved.

<u>Lemma 2.</u> If M_i are locally transversally interesecting PL submanifolds of P then $M_{J'}$ have compatible normal blockbundles.

<u>Proof.</u> Triangulate so that every inclusion $M_{J'} \subset M_{J''}$ is an inclusion of a full subcomplex, and use the standard dual cell construction.

We note that if N is blocktransversal to $M_{J'}$ with respect to the compatible normal bundles, then it follows that N is block transversal to $M_{J''}$ in a neighborhood of $M_{J'}$, $J'' \subset J'$ since the blocks are the same. Hence it follows that

<u>Lemma 3.</u> Let P be a PL manifold, $M_1, M_2, \ldots, M_s$ locally transversally intersecting submanifolds, and N another submanifold, everything

assumed to be PL. Then N can be moved by an ambient ε-isotopy which is the identity outside an ε-neighborhood of N, to be simultaneously transverse to all $M_{J'}$ with respect to a compatible system of normal blockbundles. If $\partial N \subset \partial P$ is already blocktransverse to $\{\nu_{J', J''}\}$ in the isotopy can be chosen to fix ∂P.

Proof. Use [25] to move N so N is blocktransverse to $M_{J'}$ then relative to a neighborhood of M_J move N so N is blocktransverse to all $M_{J-\{i\}}$, $i \in J$ etc. In case we assume blocktransversality in the boundary we may do this relative to the boundary. If we want ε-isotopy we can obtain that by using a very fine triangulation to define the blockbundles.

We now consider variations in PL structure.

Lemma 4. Let U be a topological manifold, M_i be PL manifolds locally flatly embedded in U. Suppose U has a PL structure U' in which M_i are PL embedded submanifolds intersecting locally transversally. Then if U is given another PL structure U" such that M_i are still PL embedded. Then they are still locally transversally intersecting if $\dim(U) \geq 6$ (≥ 5 if $\partial U = \emptyset$), and

$$\dim(U) - \dim(M_i) \geq 3 \quad \text{for all } i.$$

Proof. Let p be a point in U, and let V be an open neighborhood in U such that V, with PL structure induced from U' is PL homeomorphic to R^n by a homeomorphism h sending M_i to subvector spaces of H^n in general position. Denote the union of these subvector spaces by K. Then K is a complex of codimension 3 in R^n. If we consider V with PL structure induced from U" then $h: V'' \to H^n$ is not PL, but restricted to $V'' \cap M_i$ it is, and by uniqueness of PL structure of H^n there is a small isotopy h^t of h such that h^1 is PL. Let Z be a regular neighborhood of p in $\bigcup M_i \cap V$, then $h|Z$ and $h^1|Z$ are close PL embeddings of a finite codimension 3 complex into H^n so by [42] there is an ambient PL isotopy of H^n moving $h|Z$ to $h^1|Z$, i.e., there is a PL homeomorphism of a neighborhood W of p in U sending $\bigcup M_i \cap W$ to K, i.e., to subvector spaces in general position.

Definition 5. Let V be a topological manifold and M_i be PL manifolds locally flatly embedded in V. Then M_i are said to be tamely transversally intersecting in V if every point p in V has a neighborhood U with a PL structure so that $U \cap M_i \subset U$ in a PL embedding. ($U \cap M_i$ has PL structure induced from M_i.) Furthermore $U \cap M_i$ are required to be locally transversally intersecting in the PL sense.

Remark. We note that the question of tame transversality is independent of the PL structure on U in codimension 3 by Lemma 4 above.

We now return to our motivating situation: given

$$\varphi_i: (D^{p_i}, \partial D^{p_i}) \times (I, 0, 1) \to (V^n, \partial V) \times (I, 0, 1)$$

$n \geq 6$, $n-p_i \geq 3$, denote $\varphi_i | D^{p_i} \times 0$ by g_i and assume

$$\varphi_i | \partial D^{p_i} \times I = g_i | \partial D^{p_i} \times 1_I .$$

Then there are small ambient isotopies h_i^t that are product isotopies when restricted to $\partial V \times I$ so that $h_i^1 \circ \varphi_i$ are tamely transversally intersecting.

Proof. To facilitate notation whenever we have found isotopies as above such that $h_i^1 \circ \varphi_i$ satisfies some condition we may as well assume this was true originally and thus denote $h_i^1 \circ \varphi_i$ by φ_i. First we take an inside collar of ∂V in V, $\partial V \times [0,1]$ and of ∂D^{p_i} in D^{p_i}, $\partial D^{p_i} \times [0,1]$, and after a small ambient isotopy we can assume that φ_i agree with the collars near the boundary, see e.g. [37]. Also we may assume φ_i is a product isotopy when restricted to $V \times [0, \varepsilon]$ and $V \times [1-\varepsilon, 1]$ for some sufficiently small ε. Having done this, we assume inductively that φ_i are tamely transversally intersecting for $i < r$, and still agreeing with collars near the boundary. We let

$$\Phi : D^{p_i} \times I \times R^{n-p_i} \to V \times I$$

be an embedding extending φ_r, such that Φ is a product embedding when restricted to $\partial V \times [0, \varepsilon] \times I$, $V \times [0, \varepsilon]$ and $V \times [1-\varepsilon, 1]$, this being a trivial extension of Lemma 4 of [38]. Denote

$$\varphi_i(D^{p_i} \times I) = L_i$$

and

$$U = \Phi(D^{p_r} \times I \times R^{n-p_r})$$

and let

$$B = L_1 \cup L_2 \cup \ldots \cup L_{r-1} .$$

Let Z be a finite subcomplex of $B \cap U$ such that

$$Z \supset B \cap \Phi(D^{p_r} \times I \times B^{n-p_r})$$

where B^{n-p_r} is the unit ball of R^{n-p_r}. Then by [42] as quoted in [38] there is an ambient ε-isotopy of $U \cap V \times 0$ which is the identity outside a compact set and moves $Z \cap V \times 0$ to a PL embedding. Extending by the identity this is an ambient isotopy of $V \times 0$ and we use the product isotopy to obtain an isotopy of $V \times I$. Our considerations about collars now assure that $Z \cap V \times 1$ is PL embedded in a neighborhood of $\partial V \times 1$ and thus by [42] may be moved to a PL embedding by an isotopy as above which is the identity near the boundary. We extend to an isotopy of $V \times I$ using a product isotopy near $V \times 1$, the identity near $V \times 0$ and tapering the ambient isotopy off in between. By this we obtain that Z is PL embedded in a neighborhood of $\partial(V \times I)$ and we then finally obtain an isotopy of $V \times I$ relative to a neighborhood of $\partial(V \times I)$ so that Z is finally PL embedded in U. All these isotopies may be chosen so small that $B - Z$ stays outside $\Phi(D^{p_r} \times I \times B^{n-p_r})$. If we now shrink the fibres of U we can thus assume that $B \cap U \subset U$ is a PL embedding.

It now follows from Lemma 4 that $L_i \cap U$ are locally transverse in the PL sense for $i < r$, and we can by Lemma 3 mover L_r by an ambient ε-isotopy as above so that $L_1, L_2, \ldots, L_r$ are transversally intersecting in the PL sense. This isotopy, as before, can be done in stages assuring productisotopy when restricted to $\partial V \times [0,\varepsilon] \times I$, etc.

§ 2. Straightening concordances.

The proof of the addendum takes a little more doing, the problem being the starting point. We need to know that a codimension 2 concordance of a disc can be straightened if the complement is simply connected. To do this all we need to know is that a concordance of a simply connected topological manifold can be straightened. So we proceed to prove

Theorem. Let V^n be a simply connected topological manifold with boundary, $n \geq 5$, if $n = 5$ assume V is a handlebody. Let $h: V \times (I, 0, 1) \to V \times (I, 0, 1)$ be a homeomorphism which restricts to the identity on $\partial V \times I \cup V \times 0$. Then there is an isotopy φ' of $V \times I$ fixing $\partial V \times \Phi \cup V \times 0$ such that $\varphi' \circ h = 1_{V \times I}$.

Proof. By [43], V is a handlebody, so let $V^{(k)}$ be a handlebody filtration relative to $\emptyset$. First we deform h to a homeomorphism which is the identity on $\overline{V - V^{(2)}}$. This is done by inductive straightening of the dual handles. First, straighten the core of the handle using [38] and then a neighborhood, using, e.g. the h-cobordism theorem. This procedure breaks down when we come to codimension 2 dual handles, i.e., the 2-skeleton. The 2-skeleton however is smoothable so by [2] we can finish off the straightening.

This makes all results on embeddings, automorphisms, and concordance spaces in Chapter 3 hold as in the PL case, with the added assumption that the ambient manifold be of dimension at least 6 and the embedded manifold be a handlebody. The only place one needs to put in information is replacing $(\widetilde{PL}_n, PL_n)$ by $(\Pi op_n, Top_n)$ using [44].

One should specifically mention the following result:

<u>Theorem.</u> Let V^n, $n \geq 5$, be a topological manifold; if $n = 5$, assume V is a handlebody. Then

$$\pi_0(C(V)) = Wh_2(\pi_1(V)) \oplus Wh_1(\pi_2(V); \pi_1(V)).$$

<u>Proof.</u> As in the PL case, prove that $\pi_0(C(V)) = \pi_0(C(V^{(3)}))$, where $V^{(3)}$ is the 3-skeleton of some handlebody decomposition. Then apply [2] to conclude that it is the same as in the Diff case so one may refer to Hatcher-Wagoner.

§3. Homotoping a map to a bundle map.

Finally, we want to extend the application of homotoping a map to a bundle map to the topological case. We succeed to the following extent.

Consider the problem: Given a map $f: V^v \to M^m$ of closed topological manifolds, when is f homotopic to a bundle map? The methods of Chapter 5 along with topological transversality [43] and the topological Lemma of Disjunction give the same results as in the PL case, assuming M is triangulable. It is the purpose of this note, which is not in the most general form, to reduce the question to that case. Let D be the total space of the normal disc bundle ν of M. Then we have a diagram

$$\begin{array}{ccc} V & \xleftarrow[i'\,\subset]{p'} & E \\ {\scriptstyle f}\downarrow & & \downarrow{\scriptstyle f'} \\ M & \xleftarrow[i\,\subset]{} & D \end{array}$$

where p is the normal bundle projection, i is the zero section, f' is the pullback of f. We note that by symmetry p' is the pullback of p over f, so E is the total space of a bundle p' over V with zero section i'. Hence E is a manifold and $f' : (E, \partial E) \to (D, \partial D)$ is a map of manifolds. Also D is triangulable, being a codimension 0 submanifold of Euclidean space, and the following theorem thus completes our goal.

Theorem: Notation as above. Assume V is simply connected $v - m \neq 3$, $m \neq 4$, $v \geq 5$. Then f' is homotopic to a bundle map if and only if f is.

Proof. The 'if' part is trivial and does not need any assumptions. Assume f is homotopic to a bundle map, let F be the homotopy

$$\begin{array}{ccc} F^*(D) & \xrightarrow{F'} & D \\ \bar{p}\downarrow & & \downarrow p \\ V \times I & \xrightarrow{F} & M \end{array}$$

Then since $F^*(D)$ is a bundle over $V \times I$ it is a product bundle

$$\begin{array}{ccc} E \times I = F^*(D) & \xrightarrow{F'} & D \\ p' \times 1_I \searrow \;\; \downarrow p & & \downarrow p \\ V \times I & \xrightarrow{F} & M \end{array}$$

so F' is a homotopy from f' to $F'|E \times 1$, but

$$\begin{array}{ccc} E \times 1 & \xrightarrow{F'|E \times 1} & D \\ p|E \times 1 \downarrow & & \downarrow p \\ V & \xrightarrow{F|V \times 1} & M \end{array}$$

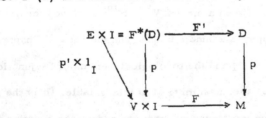

can also be considered a pullback over p, so if $F|V \times 1$ is a bundle map, then $F'|E \times 1$ is also.

Now assume f' is homotopic to a bundle map, and let F' be the homotopy $F': E \times I \to D$. $F'|E \times 0$ is f' so is clearly transverse to $M \subset D$, and $F'|E \times 1$ is also transverse to $M \subset D$ by the following argument: Since $i \circ p$ is homotopic to 1_D, $F'|E \times 1$ being a bundle map can be identified with the pullback of

$$\begin{array}{ccc} E \times 1 & \longrightarrow & i^*(E \times 1) \\ \Big\downarrow {\scriptstyle F'|E \times 1} & & \Big\downarrow \\ D & \xrightarrow{\ p\ } & M \end{array}$$

but p is a disc bundle projection, so considering the pullback diagram the other way around $E \times 1 \to i^*(E \times 1)$ is a disc bundle, and $F'|E \times 1$ is a disc bundle map, and that is what it means to be transverse to $M \subset D$. Since $v\text{-}m$ is assumed to be $\neq 3$ and $m \neq 4$, F' is homotopic relative to $E \times 0 \cup E \times 1$ to a map F which is transverse regular to M. So let $W = F^{-1}(M)$. Then W is a cobordism from V to the total space of $i^*(F|E \times 1)$ which we denote V' and we have a diagram which is homotopy commutative

$$\begin{array}{ccccc} V & \xleftarrow{\ p'\ } & E = E & \supset & V' \\ \Big\downarrow {\scriptstyle f} & & {\scriptstyle f'} \searrow \ \swarrow {\scriptstyle F|E \times 1} & & \Big\downarrow \\ M & \xleftarrow{\ p\ } & D & \supset & M \end{array}$$

and $V' \to M$ is a bundle map, so all we need to prove is that there is a homeomorphism $h: V \to V'$ so that the composition $V \xrightarrow{h} V' \subset E \xrightarrow{p'} V$ is homotopic to the identity. To prove this we set up a surgery problem. We have

$$(W, V, V') \subset (E \times I, E \times 0, E \times 1) \xrightarrow{p' \times 1} (V \times I, V \times 0, V \times 1)$$

is a degree 1 map and the restriction to $\partial W = V \cup V'$ is a homotopy equivalence.

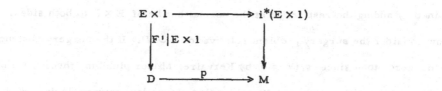

Let the normal bundle of W in $E \times I$ be denoted by ξ, then ξ is the pullback of ν over F by transversality, so the above map is covered by a bundle map

$$\begin{array}{ccc} \xi & \longrightarrow & (\xi \mid V \times 0) \times I \\ \downarrow & & \downarrow \\ W & \longrightarrow & V \times I \end{array}$$

We want the normal bundle of W over W rather than ξ, but this is obtained by adding the restriction of the normal bundle of $E \times I$ to both sides. We now consider the surgery problem relative to $V \cup V'$. If the surgery obstruction is nonzero, then since $\pi_1(V) = 0$ by Kervaire, Milnor plumbing theory, we may add a problem over a sphere with minus this obtruction, thus replacing W by a cobordism W' with trivial surgery obstruction. We may then complete surgery to obtain a cobordism W" and a homotopy equivalence

$$g: (W", V, V") \to (V \times I, V \times 0, V \times 1)$$

where $g|V = 1_V$ and $g|V'$ is the composition $V' \subset E \xrightarrow{p'} V$. It follows that W" is an h-cobordism so there is a homeomorphism

$$g': (V \times I, V \times 0, V \times 1) \to (W", V, V')$$

which is the identity on $V \times 0$. We let $h = g'|V \times 1$. Then $g \circ g'$ is a homotopy from 1_V to

$$V \xrightarrow{h} V' \subset E \xrightarrow{p'} V$$

q.e.d.

We therefore have that all the results in section 5 pertaining to the PL category also hold in the topological category for deforming $f: V \to M$ to a bundle projection, provided $\pi_1(V) = 0$.

Bibliography

1. Antonelli, Burghelea and Kahn, The Concordance Homotopy Groups of Geometric Automorphism Groups, Springer Lecture Notes, no. 215.

2. Burghelea and Lashof, The homotopy type of the space of diffeomorphisms, Part I, Trans. AMS 196(1975), 1-36.

3. Ibid, Part II, 37-50.

4. Cerf, J., Le stratification naturelle des espaces des fonctions différentielles réeles et naturelles des théorèmes de pseudo isotopie, IHES, vol. 39.

5. Hirsch, M. and Mazur, B., Smoothing of piecewise linear manifolds, Mimeo., Cambridge Univ. 1964.

6. Hudson, J. F. P., Concordance and isotopy, Ann. of Math. 91(1970), 425-448.

7. ________, Piecewise Linear Topology, Benjamin, New York, 1969.

8. Lashof and Rothenberg, Micorbundles and smoothing, Topology 3 (1965), 357-388.

9. Morlet, C., Cours Pecot, Mimeo. 1969.

10. ________, Topologie des variétes semi-linéaires, Ann. Scient. Ec. Norm. Sup., 4 ème, t 1, 1968, 313-394.

11. Rourke, C. P., Embedded Handle Theory, Concordance and Isotopy, Topology of Manifolds (Georgia Conference), Cantrell and Edwards, Editors, Markham 1969.

12. Stone, David, Stratified Polyhedra, Lecture Notes in Math, no. 252, Springer, Berlin.

13. Volodin, Z. A., Algebraic K-Theory (In Russian), Uspecki Mat. Nauk 27(1972), 207-208.

14. Wagoner, J., Algebraic invariants for pseudo-isotopies, Proc. of Liverpool Symposium II, Lecture Notes in Math., Springer-Verlag 209.

15. Wall, C.T.C., Differential Topology, Part IV, Mimeo., Cambridge.

16. Casson, A., Fibrations over spheres, Topology 6 (1967), 489-500.

17. Farrel, F.T., The obstruction to fibering a manifold over a circle, Bull. AMS 73(1967), 737-740.

18. Wall, C.T.C., Surgery on Compact Manifolds, Academic Press, New York.

19. Brumfiel, G., On the homotopy groups of BPL and PL/O, Ann. of Math. 88 (1968), 291-311.

20. Hilton, P.J., On the homotopy groups of the union of spheres, J. London Math. Soc., 30 (1955), 154-172.

21. Hatcher, A and Wagoner, J., Pseudo-isotopies of compact manifolds, Astérisque 6, Société Mathématique de France 1973.

22. Quinn, F.S., A geometric formulation of surgery, Ph.D. Thesis, Princeton University, 1969.

23. Wall, C.T.C., Finiteness conditions for CW complexes, I, Ann. of Math. 81(1965), 56-69, II Proc. Roy. Soc. A 295 (1966), 129-139.

24. Milnor, J., Whitehead torsion, Bull. AMS 72(1966), 358-426.

25. Rourke, C.P. and Sanderson, B.J., Block bundles I, Ann. of Math. 87(1968), 1-28, II ibid, 256-278, III ibid. 431-483.

26. Kuiper, N. and Lashof, R., Microbundles and bundles I, Invent. Math. 1 (1966), 1-17, II ibid. 243-259.

27. Cerf, J., Topologie de certains espaces de plongement, Bull. Soc. Math. France 89 (1961), 227-380.

28. Rourke and Sanderson, Δ-sets, I, Quart. J. of Math., Oxford 22 (1971), 321-38, II Ibid. 465-85.

29. Sullivan, D., Triangulating Homotopy Equivalences, Notes - Warwick University, 1966.

30. Sullivan, D., Smoothing Homotopy Equivalences, Notes -- Warwick University, 1966.

31. Mundres, J., Elementary Differential Topology, Princeton 1963.

32. Thom, R., Les classes caractéristiques de Pontrjagin des variétés triangulées, Internat. Symp. Algebraic Topology, Mexico 1958, 54-67.

33. Putz, H., Triangulation of fibre bundles, Canad. J. Math. 19 (1967), 499-513.

34. Palais, R., Local triviality of the restriction map for embeddings, Comm. Math. Helv. 34 (1960), 305-312.

35. Edwards and Kirby, Deformations of spaces of embeddings, Ann. of Math. 93(1971), 63-88.

36. Lees, J., Immersions and surgeries of topological manifolds, Bull. AMS 75 (1969), 529-34.

37. Rushing, T. B., Topological Embeddings, Academic Press 1973, New York.

38. Pedersen, E., Bull AMS 80 (1974), 658-660, and to appear.

39. Haefliger, A., Plongements différentiables de variétés dans variétés, Comm. Math. Helv. 36 (1961), 47-82.

40. Hudson, J., On transversality, Proc. Camb. Phil. Soc. 66 (1969), 17-20.

41. Rourke, C.P. and Sanderson, B. J., Decompositions and the relative tubular neighborhood conjecture, Topology 9 (1970), 225-229.

42. Connelly, R., Unknotting close embeddings of polyhedra, to appear. See also Proc. of the Univ. of Georgia Conference on Topology of Manifolds 1969, 384-389.

43. Kirby, R., and Siebenmann, L., On the triangulation of manifolds and the Hauptvermutung, Bull. AMS 75(1969), 742-749.

44. Rourke and Sanderson, On topological neighborhoods, Compositio Math. 22 (1970), 387-424.

45. Haefliger, A. and Poenaro, V., La classification des immersions combinatoires, Publ. Math. IHES 23 (1964), 75-91.

46. Millet, Kenneth C., Piecewise linear concordances and isotopies, Memoir 153, Am. Math. Soc. (1974).

47. Casson, A. and Gottlieb, D., Fibrations with compact fibres, to appear.

Vol. 309: D. H. Sattinger, Topics in Stability and Bifurcation Theory. VI, 190 pages. 1973. DM 20,-

Vol. 310: B. Iversen, Generic Local Structure of the Morphisms in Commutative Algebra. IV, 108 pages. 1973. DM 18,-

Vol. 311: Conference on Commutative Algebra. Edited by J. W Brewer and E. A. Rutter. VII, 251 pages. 1973. DM 24,-

Vol. 312: Symposium on Ordinary Differential Equations Edited by W A Harris, Jr. and Y Sibuya. VIII, 204 pages 1973. DM 22,-

Vol 313: K. Jörgens and J. Weidmann, Spectral Properties of Hamiltonian Operators. III, 140 pages. 1973 DM 18,-

Vol. 314: M. Deuring, Lectures on the Theory of Algebraic Functions of One Variable. VI, 151 pages 1973 DM 18,-

Vol 315: K Bichteler, Integration Theory (with Special Attention to Vector Measures). VI, 357 pages. 1973 DM 29,-

Vol 316: Symposium on Non-Well-Posed Problems and Logarithmic Convexity. Edited by R J Knops V, 176 pages. 1973. DM 20,-

Vol. 317: Séminaire Bourbaki - vol. 1971/72. Exposés 400-417 IV, 361 pages. 1973. DM 29,-

Vol. 318: Recent Advances in Topological Dynamics Edited by A Beck. VIII, 285 pages. 1973. DM 27,-

Vol. 319: Conference on Group Theory. Edited by R. W Gatterdam and K. W. Weston. V, 188 pages. 1973. DM 20,-

Vol. 320: Modular Functions of One Variable I Edited by W. Kuyk V, 195 pages. 1973. DM 20,-

Vol 321: Séminaire de Probabilités VII. Edité par P. A. Meyer. VI, 322 pages. 1973. DM 29,-

Vol 322: Nonlinear Problems in the Physical Sciences and Biology Edited by I Stakgold, D. D. Joseph and D. H. Sattinger. VIII, 357 pages 1973 DM 29,-

Vol. 323: J. L. Lions, Perturbations Singulières dans les Problèmes aux Limites et en Contrôle Optimal. XII, 645 pages. 1973 DM 46,-

Vol 324: K. Kreith, Oscillation Theory. VI, 109 pages 1973 DM 18,-

Vol. 325: C-C Chou, La Transformation de Fourier Complexe et L'Equation de Convolution IX, 137 pages 1973 DM18,-

Vol. 326: A. Robert, Elliptic Curves. VIII, 264 pages. 1973 DM 24,-

Vol. 327: E. Matlis, One-Dimensional Cohen-Macaulay Rings. XII, 157 pages 1973. DM 20,-

Vol. 328: J. R Büchi and D. Siefkes, The Monadic Second Order Theory of All Countable Ordinals VI, 217 pages 1973. DM 22,-

Vol. 329: W. Trebels, Multipliers for (C, α)-Bounded Fourier Expansions in Banach Spaces and Approximation Theory. VII, 103 pages. 1973. DM 18,-

Vol. 330: Proceedings of the Second Japan-USSR Symposium on Probability Theory. Edited by G. Maruyama and Yu. V. Prokhorov VI, 550 pages. 1973 DM 40,-

Vol. 331: Summer School on Topological Vector Spaces Edited by L Waelbroeck. VI, 226 pages. 1973. DM 22,-

Vol. 332: Séminaire Pierre Lelong (Analyse) Année 1971-1972. V, 131 pages. 1973. DM 18,-

Vol. 333: Numerische, insbesondere approximationstheoretische Behandlung von Funktionalgleichungen. Herausgegeben von R. Ansorge und W Törnig VI, 296 Seiten. 1973. DM 27,-

Vol. 334: F. Schweiger, The Metrical Theory of Jacobi-Perron Algorithm. V, 111 pages. 1973. DM 18,-

Vol. 335: H Huck, R. Roitzsch, U. Simon, W. Vortisch, R. Walden, B. Wegner und W. Wendland, Beweismethoden der Differentialgeometrie im Großen. IX, 159 Seiten. 1973 DM 20,-

Vol 336: L'Analyse Harmonique dans le Domaine Complexe Edité par E. J Akutowicz. VIII, 169 pages 1973. DM 20,-

Vol. 337: Cambridge Summer School in Mathematical Logic Edited by A. R. D. Mathias and H. Rogers. IX, 660 pages. 1973. DM 46,-

Vol. 338: J Lindenstrauss and L Tzafriri, Classical Banach Spaces IX, 243 pages 1973 DM 24,-

Vol. 339: G. Kempf, F. Knudsen, D. Mumford and B. Saint-Donat, Toroidal Embeddings I. VIII, 209 pages. 1973. DM 22,-

Vol. 340: Groupes de Monodromie en Géométrie Algébrique. (SGA 7 II). Par P. Deligne et N. Katz. X, 438 pages. 1973. DM 44,-

Vol. 341: Algebraic K-Theory I, Higher K-Theories. Edited by H. Bass. XV, 335 pages. 1973. DM 29,-

Vol. 342: Algebraic K-Theory II, "Classical" Algebraic K-Theory, and Connections with Arithmetic. Edited by H. Bass. XV, 527 pages. 1973. DM 40,-

Vol. 343: Algebraic K-Theory III, Hermitian K-Theory and Geometric Applications. Edited by H. Bass. XV, 572 pages. 1973. DM 40,-

Vol. 344: A. S. Troelstra (Editor), Metamathematical Investigation of Intuitionistic Arithmetic and Analysis. XVII, 485 pages. 1973. DM 38,-

Vol 345: Proceedings of a Conference on Operator Theory Edited by P A Fillmore. VI, 228 pages 1973 DM 22,-

Vol. 346: Fučík et al., Spectral Analysis of Nonlinear Operators. II, 287 pages 1973 DM 26,-

Vol. 347: J. M. Boardman and R. M. Vogt, Homotopy Invariant Algebraic Structures on Topological Spaces. X, 257 pages. 1973. DM 24,-

Vol. 348: A. M. Mathai and R. K. Saxena, Generalized Hypergeometric Functions with Applications in Statistics and Physical Sciences. VII, 314 pages. 1973. DM 26,-

Vol. 349: Modular Functions of One Variable II. Edited by W. Kuyk and P. Deligne. V, 598 pages. 1973. DM 38,-

Vol. 350: Modular Functions of One Variable III. Edited by W. Kuyk and J.-P. Serre. V, 350 pages. 1973. DM 26,-

Vol. 351: H. Tachikawa, Quasi-Frobenius Rings and Generalizations. XI, 172 pages. 1973. DM 20,-

Vol 352: J D. Fay, Theta Functions on Riemann Surfaces. V, 137 pages. 1973. DM 18,-

Vol 353: Proceedings of the Conference on Orders, Group Rings and Related Topics. Organized by J S Hsia, M. L. Madan and T. G Ralley. X, 224 pages. 1973. DM 22,-

Vol 354: K J. Devlin, Aspects of Constructibility. XII, 240 pages. 1973 DM 24,-

Vol. 355: M. Sion, A Theory of Semigroup Valued Measures. V, 140 pages. 1973. DM 18,-

Vol. 356: W. L. J. van der Kallen, Infinitesimally Central-Extensions of Chevalley Groups. VII, 147 pages. 1973. DM 18,-

Vol. 357: W. Borho, P. Gabriel und R. Rentschler, Primideale in Einhüllenden auflösbarer Lie-Algebren. V, 182 Seiten 1973. DM 20,-

Vol. 358: F. L. Williams, Tensor Products of Principal Series Representations. VI, 132 pages. 1973. DM 18,-

Vol 359: U Stammbach, Homology in Group Theory. VIII, 183 pages 1973 DM 20,-

Vol. 360: W. J. Padgett and R. L. Taylor, Laws of Large Numbers for Normed Linear Spaces and Certain Fréchet Spaces. VI, 111 pages 1973 DM 18,-

Vol. 361: J. W. Schutz, Foundations of Special Relativity: Kinematic Axioms for Minkowski Space Time. XX, 314 pages. 1973. DM 26,-

Vol. 362: Proceedings of the Conference on Numerical Solution of Ordinary Differential Equations Edited by D. Bettis VIII, 490 pages 1974 DM 34,-

Vol. 363: Conference on the Numerical Solution of Differential Equations Edited by G. A Watson IX, 221 pages. 1974 DM 20,-

Vol 364: Proceedings on Infinite Dimensional Holomorphy. Edited by T. L. Hayden and T. J. Suffridge. VII, 212 pages. 1974. DM 20,-

Vol. 365: R. P. Gilbert, Constructive Methods for Elliptic Equations. VII, 397 pages. 1974. DM 26,-

Vol. 366: R. Steinberg, Conjugacy Classes in Algebraic Groups (Notes by V. V. Deodhar). VI, 159 pages. 1974. DM 18,-

Vol. 367: K. Langmann und W. Lütkebohmert, Cousinverteilungen und Fortsetzungssätze. VI, 151 Seiten. 1974. DM 16,-

Vol. 368: R. J. Milgram, Unstable Homotopy from the Stable Point of View. V, 109 pages. 1974. DM 16,-

Vol. 369: Victoria Symposium on Nonstandard Analysis Edited by A. Hurd and P Loeb XVIII, 339 pages. 1974 DM 26,-

Vol. 370: B. Mazur and W. Messing, Universal Extensions and One Dimensional Crystalline Cohomology VII, 134 pages. 1974 DM 16,-

Vol 371: V Poenaru, Analyse Différentielle V, 228 pages 1974 DM 20,-

Vol. 372: Proceedings of the Second International Conference on the Theory of Groups 1973. Edited by M. F. Newman. VII, 740 pages. 1974. DM 48,-

Vol 373: A E R. Woodcock and T. Poston, A Geometrical Study of the Elementary Catastrophes. V, 257 pages 1974 DM 22,-

Vol. 374: S Yamamuro, Differential Calculus in Topological Linear Spaces IV, 179 pages. 1974. DM 18,-

Vol 375: Topology Conference 1973 Edited by R. F. Dickman Jr. and P. Fletcher. X, 283 pages 1974. DM 24,-

Vol. 376: D. B. Osteyee and I J Good, Information, Weight of Evidence, the Singularity between Probability Measures and Signal Detection XI, 156 pages. 1974 DM 16 -

Vol. 377: A. M. Fink, Almost Periodic Differential Equations VIII, 336 pages. 1974. DM 26,-

Vol. 378: TOPO 72 - General Topology and its Applications. Proceedings 1972. Edited by R. Alò, R. W. Heath and J Nagata XIV, 651 pages. 1974. DM 50,-

Vol. 379: A. Badrikian et S Chevet, Mesures Cylindriques, Espaces de Wiener et Fonctions Aléatoires Gaussiennes. X, 383 pages. 1974. DM 32,-

Vol. 380: M. Petrich, Rings and Semigroups. VIII, 182 pages. 1974. DM 18,-

Vol. 381: Séminaire de Probabilités VIII. Edité par P. A. Meyer. IX, 354 pages. 1974. DM 32,-

Vol 382: J H van Lint, Combinatorial Theory Seminar Eindhoven University of Technology VI, 131 pages 1974 DM 18,-

Vol. 383: Séminaire Bourbaki - vol. 1972/73 Exposés 418-435 IV, 334 pages. 1974. DM 30,-

Vol. 384: Functional Analysis and Applications, Proceedings 1972 Edited by L. Nachbin. V, 270 pages 1974. DM 22,-

Vol. 385: J. Douglas Jr. and T. Dupont, Collocation Methods for Parabolic Equations in a Single Space Variable (Based on C^1-Piecewise-Polynomial Spaces). V, 147 pages. 1974. DM 16,-

Vol. 386: J. Tits, Buildings of Spherical Type and Finite BN-Pairs. IX, 299 pages. 1974. DM 24,-

Vol 387: C. P Bruter, Eléments de la Théorie des Matroïdes. V, 138 pages. 1974. DM 18,-

Vol. 388: R. L Lipsman, Group Representations X, 166 pages. 1974. DM 20,-

Vol. 389: M.-A. Knus et M. Ojanguren, Théorie de la Descente et Algèbres d' Azumaya. IV, 163 pages. 1974. DM 20,-

Vol 390: P A. Meyer, P. Priouret et F. Spitzer, Ecole d'Eté de Probabilités de Saint-Flour III - 1973. Edité par A. Badrikian et P.-L. Hennequin. VIII, 189 pages 1974. DM 20,-

Vol. 391: J. Gray, Formal Category Theory: Adjointness for 2-Categories XII, 282 pages. 1974 DM 24,-

Vol 392: Géométrie Différentielle, Colloque, Santiago de Compostela, Espagne 1972 Edité par E Vidal VI, 225 pages 1974. DM 20,-

Vol. 393: G. Wassermann, Stability of Unfoldings. IX, 164 pages 1974. DM 20,-

Vol. 394: W. M. Patterson 3rd, Iterative Methods for the Solution of a Linear Operator Equation in Hilbert Space - A Survey. III, 183 pages 1974 DM 20,-

Vol 395: Numerische Behandlung nichtlinearer Integrodifferential- und Differentialgleichungen Tagung 1973 Herausgegeben von R Ansorge und W Törnig VII, 313 Seiten 1974 DM 28,-

Vol 396: K. H. Hofmann, M. Mislove and A. Stralka, The Pontryagin Duality of Compact O-Dimensional Semilattices and its Applications. XVI, 122 pages. 1974 DM 18,-

Vol 397: T. Yamada, The Schur Subgroup of the Brauer Group V, 159 pages. 1974. DM 18,-

Vol. 398: Théories de l'Information, Actes des Rencontres de Marseille-Luminy, 1973. Edité par J Kampé de Fériet et C. Picard XII, 201 pages. 1974. DM 23,-

Vol. 399: Functional Analysis and its Applications, Proceedings 1973. Edited by H. G. Garnir, K. R. Unni and J. H. Williamson. XVII, 569 pages. 1974. DM 44,-

Vol. 400: A Crash Course on Kleinian Groups - San Francisco 1974. Edited by L. Bers and I Kra. VII, 130 pages. 1974 DM 18,-

Vol 401: F. Atiyah, Elliptic Operators and Compact Groups. V, 93 pages 1974 DM 18,-

Vol. 402: M. Waldschmidt, Nombres Transcendants VIII, 277 pages 1974. DM 25,-

Vol. 403: Combinatorial Mathematics - Proceedings 1972. Edited by D. A. Holton. VIII, 148 pages. 1974. DM 18,-

Vol. 404: Théorie du Potentiel et Analyse Harmonique Edité par J Faraut. V, 245 pages. 1974. DM 25,-

Vol. 405: K. Devlin and H. Johnsbråten, The Souslin Problem. VIII, 132 pages. 1974. DM 18,-

Vol. 406: Graphs and Combinatorics - Proceedings 1973. Edited by R. A. Bari and F. Harary. VIII, 355 pages. 1974. DM 30,-

Vol. 407: P. Berthelot, Cohomologie Cristalline des Schémas de Caracteristique $p > o$. VIII, 598 pages. 1974. DM 44,-

Vol. 408: J. Wermer, Potential Theory. VIII, 146 pages. 1974. DM 18,-

Vol. 409: Fonctions de Plusieurs Variables Complexes, Séminaire François Norguet 1970-1973. XIII, 612 pages. 1974. DM 47,-

Vol. 410: Séminaire Pierre Lelong (Analyse) Année 1972-1973. VI, 181 pages. 1974. DM 18,-

Vol. 411: Hypergraph Seminar Ohio State University, 1972. Edited by C Berge and D. Ray-Chaudhuri. IX, 287 pages. 1974 DM 28,-

Vol 412: Classification of Algebraic Varieties and Compact Complex Manifolds. Proceedings 1974 Edited by H. Popp. V, 333 pages. 1974. DM 30,-

Vol 413: M Bruneau, Variation Totale d'une Fonction. XIV, 332 pages. 1974 DM 30,-

Vol 414: T. Kambayashi, M Miyanishi and M Takeuchi, Unipotent Algebraic Groups VI, 165 pages 1974 DM 20,-

Vol. 415: Ordinary and Partial Differential Equations, Proceedings of the Conference held at Dundee, 1974 XVII, 447 pages. 1974. DM 37,-

Vol 416: M E Taylor, Pseudo Differential Operators. IV, 155 pages. 1974 DM 18,-

Vol. 417: H. H. Keller, Differential Calculus in Locally Convex Spaces XVI, 131 pages 1974 DM 18,-

Vol 418: Localization in Group Theory and Homotopy Theory and Related Topics Battelle Seattle 1974 Seminar. Edited by P J Hilton VI, 171 pages. 1974. DM 20,-

Vol 419: Topics in Analysis - Proceedings 1970. Edited by O. E. Lehto, I. S. Louhivaara, and R H Nevanlinna. XIII, 391 pages 1974. DM 35,-

Vol 420: Category Seminar. Proceedings, Sydney Category Theory Seminar 1972/73. Edited by G M Kelly VI, 375 pages 1974 DM 32,-

Vol. 421: V Poénaru, Groupes Discrets. VI, 216 pages 1974 DM 23,-

Vol 422: J.-M Lemaire, Algèbres Connexes et Homologie des Espaces de Lacets. XIV, 133 pages. 1974. DM 23,-

Vol. 423: S. S. Abhyankar and A M. Sathaye, Geometric Theory of Algebraic Space Curves. XIV, 302 pages 1974. DM 28,-

Vol 424: L Weiss and J. Wolfowitz, Maximum Probability Estimators and Related Topics. V, 106 pages. 1974. DM 18,-

Vol 425: P. R. Chernoff and J. E. Marsden, Properties of Infinite Dimensional Hamiltonian Systems. IV, 160 pages 1974 DM 20,-

Vol. 426: M L Silverstein, Symmetric Markov Processes IX, 287 pages. 1974. DM 28,-

Vol. 427: H Omori, Infinite Dimensional Lie Transformation Groups XII, 149 pages. 1974 DM 18,-

Vol 428: Algebraic and Geometrical Methods in Topology, Proceedings 1973. Edited by L. F. McAuley. XI, 280 pages 1974. DM 28,-